AF780086

Brain Theory From A Circuits And Systems Perspective

Springer Series in Cognitive and Neural Systems

Volume 6

Series Editor
Vassilis Cutsuridis
Boston University, Boston, MA, USA

For further volumes:
http://www.springer.com/series/8572

John Robert Burger

Brain Theory From A Circuits And Systems Perspective

How Electrical Science Explains Neuro-circuits, Neuro-systems, and Qubits

by John Robert Burger, PhD (UCLA)

Dr. Burger, Professor Emeritus, California State University Northridge

Springer

John Robert Burger
Veneta, Oregon, USA

ISBN 978-1-4614-6411-2 ISBN 978-1-4614-6412-9 (eBook)
DOI 10.1007/978-1-4614-6412-9
Springer New York Heidelberg Dordrecht London

Library of Congress Control Number: 2013931948

© Springer Science+Business Media New York 2013
This work is subject to copyright. All rights are reserved by the Publisher, whether the whole or part of the material is concerned, specifically the rights of translation, reprinting, reuse of illustrations, recitation, broadcasting, reproduction on microfilms or in any other physical way, and transmission or information storage and retrieval, electronic adaptation, computer software, or by similar or dissimilar methodology now known or hereafter developed. Exempted from this legal reservation are brief excerpts in connection with reviews or scholarly analysis or material supplied specifically for the purpose of being entered and executed on a computer system, for exclusive use by the purchaser of the work. Duplication of this publication or parts thereof is permitted only under the provisions of the Copyright Law of the Publisher's location, in its current version, and permission for use must always be obtained from Springer. Permissions for use may be obtained through RightsLink at the Copyright Clearance Center. Violations are liable to prosecution under the respective Copyright Law.
The use of general descriptive names, registered names, trademarks, service marks, etc. in this publication does not imply, even in the absence of a specific statement, that such names are exempt from the relevant protective laws and regulations and therefore free for general use.
While the advice and information in this book are believed to be true and accurate at the date of publication, neither the authors nor the editors nor the publisher can accept any legal responsibility for any errors or omissions that may be made. The publisher makes no warranty, express or implied, with respect to the material contained herein.

Printed on acid-free paper

Springer is part of Springer Science+Business Media (www.springer.com)

Preface

This book contributes a circuits and systems approach that emphasizes the authority of electrical principles. Mostly this book is founded on generally accepted properties of neurons, which have been well studied and are sufficient to enable a circuit model for an idealized neuron. This model in turn permits the simulation of neural logic in its various forms. Circuit simulations, executed objectively and transparently using software that is proven to function correctly, such as WinSpice as used for the experiments attached to this book, suggest important neural subcircuits that engineers and neuroscientists should find quite interesting. As a general benefit to all concerned, these subcircuits suggest a plausible new system for consciousness that would not be evident within the limits of molecular biology and biochemistry.

Beyond molecules, a brain is obviously a working complexity of circuits. There is no reason to suspect that a brain violates even the smallest aspect of circuit theory or, for that matter, any known principle of physics. So when explaining a brain, circuits and systems theory is, without question, most relevant.

Engineers often point to computerized chess, supercomputer calculations and useful robots that clean the floor as evidence that we are on the road to intelligent machines. But such achievements are largely expressions of human intelligence, not necessarily examples of intrinsic machine intelligence. The computer as an intelligent entity has a long way to go to match human intelligence. To light the way, it is inspiring to study, or at least glimpse reasonable conjectures about the brain as a system of physical neurons.

In particular, the human brain is modeled below as a system of pulsating neurons, complete with electrical connections to other pulsating neurons. To demonstrate what this system does, it was necessary to model synapses and memory in a more detailed way, employing new and exciting circuit elements without which a brain could not possibly function as a system. The result is a proposed new system for human stream of consciousness.

From an academic point of view, this book introduces an exciting new field of *Neuro circuits and Neuro systems* (NCANS). In this field, proposed circuit elements and proposed systems are introduced and tried in order to explain a brain, and to help formulate meaningful scientific experimentation. Explaining a brain is extremely important since more often than not, we are left with no clear explanations of exactly how stream of consciousness works physically.

Neuro-circuits and Neuro-systems is definitely outside the box of popular engineering areas such as circuits and systems (CAS) theory, artificial neural networks (ANN), and others that normally fall under and tend to focus on hardware systems. The emphasis below is on neural pulses, generation, and coordination, which is very different from AC/DC electronics. Shown below are neuro-synaptic combinations that result in novel implementations of logic, such as dendritic logic and enabled logic, plus a variety of unique timing elements, such as weak synapses and short-term memory neurons, that should be interesting to circuit theorists as well as to neuroscientists and neural network engineers.

A vital question is: How does one duplicate the right choices in artistic appraisal, common sense, truth judgment, understanding, and other hallmarks of human ability? There seems to be unfathomed layers of computational power within the brain not anticipated by today's technology. In view of this observation, new radical concepts continue to be entertained.

Quantum computers and qubits are examples of radical concepts that go far beyond logic circuitry. A qubit, as shown in this book, can be simulated in part by a recursive neuron, that is, one that feeds back on itself. Recursive neurons, appropriately configured, are herein termed *simulated qubits*; they are entirely classical and are of interest biologically and for engineering purposes. For instance, simulated qubits can hold true and false simultaneously, each with a specified probability. Simulated qubits are better than ordinary neurons for certain purposes.

A major advantage of simulated qubits is that they are easily configured into controlled toggle devices. Controlled toggles, when properly developed, constitute a major source of computational power, and may very well underlie consciousness and perhaps the gifted abilities of savants. Controlled toggles may be massively parallel and with the would-be advantages of reversible computing. They may someday be found to be implemented by molecules within neurons, while truly quantum systems are less likely to exist due to thermal activity (and decoherence) at body temperatures.

This book carries a reader to the edge of scientific knowledge by introducing *neuroquantology*, which is a studied mix of neuroscience, quantum physics, and psychological phenomena. The discovery of physical qubits within brain cells would be revolutionary, so much so that reasonable people cannot ignore this possibility. Qubits within a quantum system hold promise for powerful quantum computing, and weird teleportation without fields or waves.

Several asynchronous subsystems are covered in this book, involving neuro-synaptic logic in a worthwhile review of digital analysis and design. Subjects introduced are cue editors, random selection of cues, associative memory design, associative memory search, associative memorization, recall priority digital calculations, and digital selection of maximum priority. Each chapter contains a few easy end-of-chapter exercises as an aid to self-education. An appendix provides an experimental lab revolving around the simulation of neural logic.

This book gives a reader a markedly better understanding of cerebral behavior, including mechanisms for brain sense and stream of consciousness. Such topics are not just discussed verbally; they are explored technically with reference to physical circuits.

Veneta, Oregon, USA John Robert Burger

Acknowledgements

Many thanks to the Department of Electrical and Computer Science at California State University, Northridge, especially Dr. Ray Pettit for his whole-hearted support.

Contents

Chapter 1
Brain Structure

Introduction

In order to go into the field of neurocircuits and neurosystems it is important to have a basic vocabulary for the physical parts of a brain. This is true even if one is mainly interested in the logical and system aspects of brain functioning, partly because there is a need to communicate with others who are not specialists in circuits and systems, but who are familiar with brain organs. Beyond a need to talk to others, it is important to appreciate the complexity of a brain with the goal of understanding how it works. In particular, specific areas seem to be involved in intelligence and consciousness, which a student ought to be aware of, such as the *cerebral cortex*, the *hippocampus*, the *amygdala*, the *cingulated gyrus*, and the *corpus callosum* [1]. It is wise to be aware of these and all other major areas.

This chapter not only introduces basic brain anatomy but also neurons, about 100 billion of them, each capable of realizing significant logic. The logical abilities of a neuron may be verified by simulation experiments as suggested in an appendix to this book. These amazing neural abilities will be built up later in this book to produce an intelligent stream of consciousness that, for self-preservation, avoids mental blocks and avoids becoming confused by multiple recalls.

Brain Structure

The four major lobes of the brain and their locations are shown in Fig. 1.1. Note that the boundaries of the lobes are not distinct but rather are named after the bones of the skull that overlie them. The *cerebrum* is the general volume of the right and left hemisphere regions in the upper forebrain. It contains the cerebral cortex, a layer of neural gray matter, and the underlying connections, or white matter. Within this mass of neurons is nature's implementation of higher intelligence and all that goes

J.R. Burger, *Brain Theory From A Circuits And Systems Perspective*, Springer Series in Cognitive and Neural Systems 6, DOI 10.1007/978-1-4614-6412-9_1,
© Springer Science+Business Media New York 2013

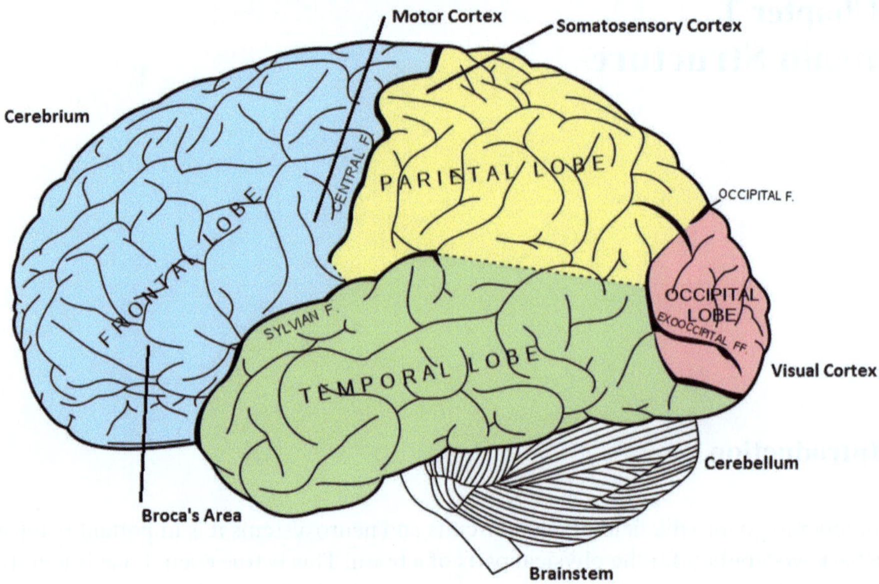

Fig. 1.1 Overview schematic of the brain (courtesy Human_Brains, Wikipedia Commons, October 2012)

with it, including what we term consciousness, often equated below to conscious short-term memory.

The cerebrum consisting of the *prefrontal cortex* and the *frontal lobe* is where this book begins. The other three lobes serve mainly to reduce sensory information to a logical format representing attributes that are usable in the frontal lobe. Time and space unfortunately do not permit analyses of the backmost three lobes. The focus in this book is on generating an intelligent stream of consciousness while leaving for another time the encoding of sensory information into attributes.

Frontal Lobe

The frontal lobe of the cerebral cortex is thought to be where important memories are activated for decisions and actions. It is in front of the ears. The human frontal lobes and the prefrontal cortex directly behind the forehead are thought to serve to provide solutions to problems, either novel or familiar, possibly generic solutions that have been worked out beforehand and stored in long-term memory. A back-to-front architecture is theorized for the cortical areas: The back area contains models of and can recognize danger; the front area stores and enables appropriate responses. Many mental disorders such as attention deficits, frontal lobe dementia, and schizophrenia are known to be caused by malfunctions in the frontal lobes.

Broca's Area

Broca's area is located in the lower back part of the left frontal lobe and is associated with the production of spoken and written language. Paul Broca in 1861 discovered the importance of this region. Broca's aphasia is a disorder in which a person can understand language but has difficulty in expressing it.

Motor Cortex

The *motor cortex* is a narrow band in the posterior part of the frontal lobes devoted to the activation of body movements, a band going from ear to ear. The motor cortex works backwards and upside down: The left side regulates muscles on the right side of the body, and the right side regulates muscles on the left side of the body. The area at the top regulates our feet, and the area at the bottom, that is, closest to our ears, regulates facial expressions. Most of the volume of the motor cortex in humans is devoted to the complex precise movements of the hands and face.

Parietal Lobes

The *parietal lobes* are in the upper back of the cerebral cortex, behind the ears, and participate in recognition of danger and opportunity. They monitor touch sensations and body positions.

Occipital Lobes

The *occipital lobes* monitor vision.

Temporal Lobes

The *temporal lobes* monitor sound.

Somatosensory Cortex

The *somatosensory cortex* is a narrow band just behind the motor cortex devoted to touch sensations from the body. The left side receives sensations from the right side of the body, and the right side receives sensations from the left side of the body. Again we are represented upside down with the top of the brain receiving sensations for our feet, and the bottom closest to our ears, receiving sensations from our face. The face and hands are highly sensitive to touch and require most of the somatosensory cortex.

Angular Gyrus

The *angular gyrus* within the temporal lobes is a left hemispheric bundle, although occasionally it is found on the right. It is a connector between a written word recognizer and the forward regions of the brain.

Planum Temporale

The *planum temporale* is associated with language comprehension. Typically it is larger in the left hemisphere within Wernicke's area.

Wernicke's Area

Wernicke's area is named after the German neurologist, Carl Wernicke; it is a language comprehension region located usually in the upper posterior left temporal lobe. Damage (lesions) in this region was found to result in serious problems of comprehension.

Cerebellum

The *cerebellum* or little brain is located behind the brainstem, near the little bump at the lower back of the head. It is thought to be a support system for cognitive functions involving sensory input. It has more neurons than other parts of the brain.

Brainstem

The *brainstem* is a centimeter or two in diameter and relays information between body and brain. The back of the brainstem is connected to the cerebellum, with a number of specialized structures to the hindbrain and the midbrain.

Interior Parts

Important interior parts of a brain are diagramed in Fig. 1.2.

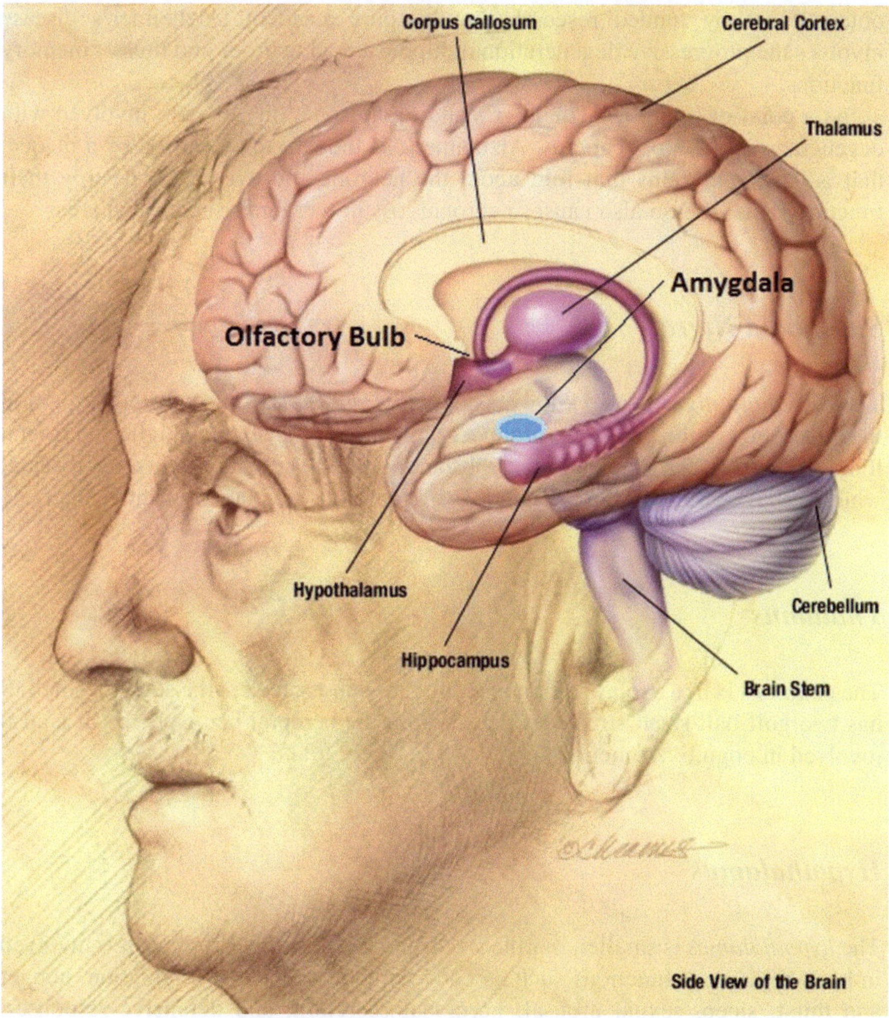

Fig. 1.2 Interior parts of the brain (courtesy Human_Brains, Wikipedia Commons, October 2012)

Hippocampus

The hippocampus is a Greek word for Neptune's seahorse, referring to its shape. The *hippocampus* refers to a pair of curved components under the inside of the temporal lobes. It plays a fundamental role in the formation and retrieval of long-term memories.

It was learned that the hippocampus is related to memory when a man, called HM, accidentally had his hippocampus and amygdala removed. He subsequently lived in a world of short-term memory only. This is an example of knowledge obtained not by funded research, but by pure accident. Alzheimer's disease involves the progressive degeneration of hippocampal neurons and thus of memory functions.

As a point of interest, Alzheimer's and Parkinson's diseases are involved with degeneracy of the hippocampus. Alzheimer's is thought to be caused by a plaque, that is, sticky proteins that fold about the neurons and eventually destroy their function. This disease also causes neuronal loss in (cerebral) subcortical areas.

Substantia Nigra

The *substantia nigra* is also known as the *black substance*. The substantia nigra in the midsection of the brainstem produces dopamine for the basal ganglia and frontal lobes. Extensive cell demise in the substantia nigra results in Parkinson's disease, which usually results in the gradual loss of conscious regulation of movement.

Thalamus

The *thalamus* is like a relay station in sorting the senses, except the sense of smell. It has two golf ball-sized structures deep within the cerebral hemispheres. It is also involved in cognitive functions.

Hypothalamus

The *hypothalamus* is smaller, marble sized, and is below the thalamus; it is involved in body regulatory functions such as body temperature, blood circulation, hunger and thirst, sleep, sexual arousal, hormonal secretion, and defensive responses. Sometimes it is called the brain's brain; it is in the center behind our eyes.

Olfactory Bulb

The *olfactory bulb* is the initial odor-processing region. Roughly 30 differing molecules are involved in approximately 10,000 different odors. Seven primary odors in ordinary terms would be peppermint, roses, nail polish, musk, camphor, vinegar, and rotten eggs. Many animals have a more powerful sense of smell than humans. The olfactory bulb plays a role in the recall of emotional memories. However, for some reason, we cannot recall the odor of an imagined object as easily as we can recall its shape and color.

Amygdala

The *amygdala* is a pair of complex parts the size of a cherry that activates primal fears and other emotional responses. It has been called the fear button. Each amygdala is about 4 cm into your brain from either temple. A sudden movement or loud sound will cause the amygdala to signal the hypothalamus to initiate a standard response, for example, to fight, to take flight, or to freeze.

The amygdala is important to memory because it adds emotional content to an impression. Emotion is an important cue for creating and for retrieving long-term memories. We usually remember details associated with emotional events, such as our first romance, or where we were when Kennedy was assassinated, or where we were on September 11, 2001.

Cingulated Gyrus

The *cingulated gyrus* aids in retrieving and analyzing memories of prior situations and developing responses for future use. It is reserved for challenges that do not require an immediate response. It helps us to make up our minds about what to do in the presence of ambiguous information. It also determines the emotional strength of sensory information and presents a result to the cortex. The cingulated gyrus is on top of the corpus callosum.

Corpus Callosum

The *corpus callosum* is a major part of a network that connects the left and right brains.

Forebrain Structure

The upper forebrain is a layer of neurons, or gray matter, and the underlying connections, or neural white matter.

Gray Matter

Gray matter refers to dendrites, which do logical processing.

White Matter

White matter refers to myelinated axons, which are insulated with myelin to transmit electrical impulses faster than what is possible in uninsulated dendrites.

Cerebrum

The cerebrum contains six layers of folded neural tissue also known as the cerebral cortex, constituting most of the neural volume of the brain.

Cerebral Cortex

The cerebral cortex is composed of *pyramidal* and *stellate* (starlike) neurons. Figure 1.3 illustrates six layers organized roughly as highly interconnected columns. Through the six layers of gray matter extend hundreds of millions of interconnected hairlike *mini columns*, each with about 100 neurons. Layer 4 is thought to be for input. Layers 5 and 6 are thought to be for output. General processing occurs in layers 1, 2, and 3. Possibly each mini column can process an attribute such as a shape or a sound. One hundred mini columns form a *macro column*, about the thickness of pencil lead.

Thousands of macro columns form 1 of the 50 or more *Brodmann areas* in each hemisphere. Korbinian Brodmann was a pioneer in organizing the cerebral cortex into distinct regions based on structural variations. He identified over 50 numbered areas in 1909, and the system is still widely used, although others have identified hundreds of additional structural variations.

The numbers add as follows. About 50 Brodmann areas might have, say, 10,000 macro columns; each macro column has perhaps 500 mini columns; each mini column might have about 100 neurons. This represents 25 billion (25×10^9) neurons per hemisphere. Each neuron may have a great many logic gates. Some undoubtedly are unused, ready to be turned to new purposes, and some are copies of others for backup purposes, but even so, the amount of logic is astounding.

Gray matter is distributed roughly toward the outside as in Fig. 1.4. White matter forms the inner region and indeed connects the left brain to the right brain.

Corpus Callosum

The hemispheres are connected by the *corpus callosum* and the *anterior commissure*. The corpus callosum, or "colossal commissure," is a wide, flat bundle of

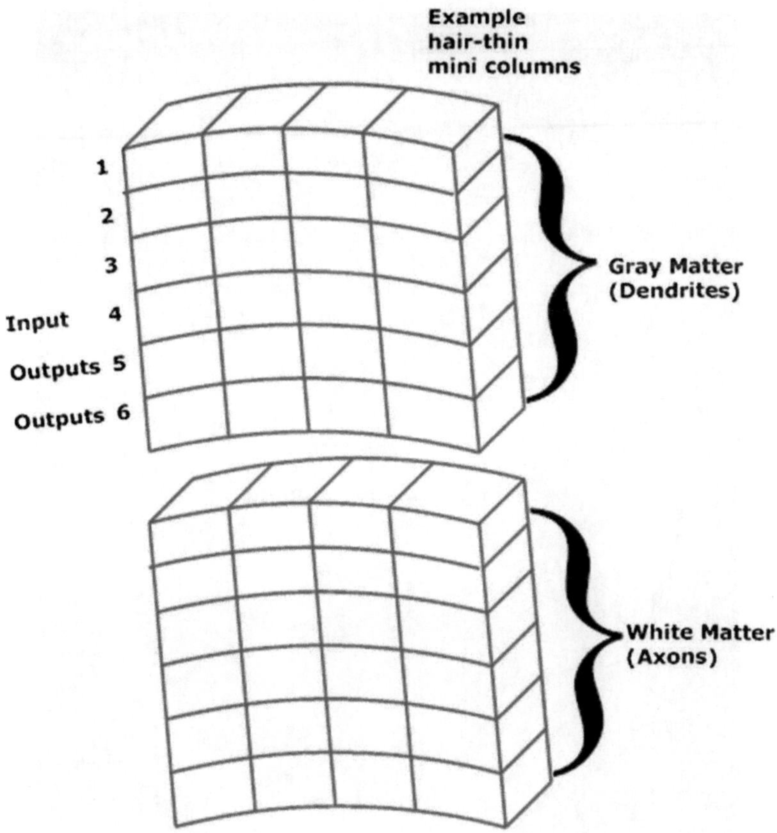

Fig. 1.3 Cerebral cortex organized into hair-thin columns

neural fibers that facilitates general interhemispheric communication. It is the largest white matter structure in the brain, consisting of 200–250 million axons. Note that the cerebrum in the forebrain refers to the combined right and left hemispheres. Again, each hemisphere is divided into four lobes: three in back to receive and analyze incoming sensory information and a frontal lobe to construct a response.

Anterior Commissure

The anterior commissure is a separate bundle to connect the two amygdala and aids in signaling sharp pain, olfaction, instinct, and sexual behavior.

Aside from redundancy for reliability, the right brain is geared to novel challenges, for instance strange faces and new language. The left brain is geared

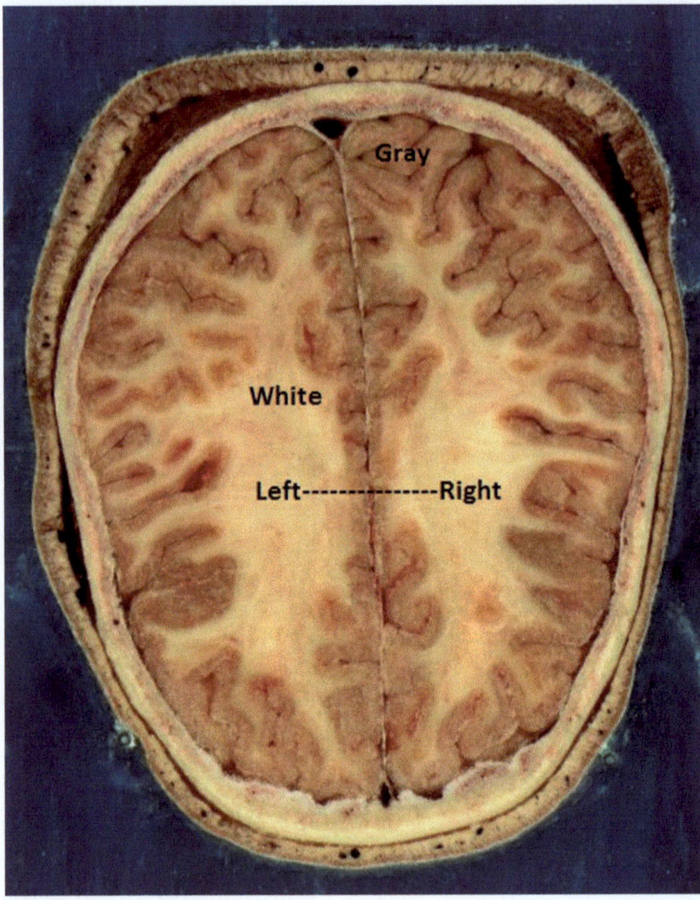

Fig. 1.4 Cross section of gray and underlying white matter (courtesy Human_Brains, Wikipedia Commons, October 2012)

to the familiar, for instance, familiar faces and regular language. It is said that a baby uses the right brain for nonverbal skills, but soon develops a language template in the left brain. Overall, the right and left hemispheres are tightly connected. If one-half becomes inoperative, the other half can take over.

Glial Cells, Neuroglia, or Glia

Glial cells, approximately one for each neuron, form *myelin*, which insulates parts of neurons, and glial cells provide support and protection for neurons in the brain, and for neurons in other parts of the nervous system.

Sensory Inputs

Sight, sound, touch, taste, and odor send over a million parallel signals via appropriate nerves to respective parts of the brain. Sensory signals are encoded into a useful set of logical signals termed *attributes*. The 1.2 million axon fibers in each eye's optic nerve, for example, are first processed in the thalamus, then in the amygdala for emotional analysis, and then within credit card-sized areas in the occipital lobes. Additional processing occurs in the cerebral cortex. As a result, the existence of numerous attributes such as edges and textures become available.

Each attribute construed from the senses has its own unique cells and its own physical region. In a *distributed model*, each memory element, for instance, is dedicated to a particular attribute such that no other attribute would ever enter it. Memory like this has been termed *distributed memory* [2] where each attribute has its own respective memory path. It is biologically efficient, but cannot be used in man-made machines since it results in excessive wiring.

The process of converting analog sensory inputs into attributes, each true or false, denoting the presence or the absence of a given feature, is termed *encoding*. The brain continuously encodes sensory information into attributes, which surely is a major undertaking. Sensory encoding is beyond the scope of this book.

Another form of encoding that is completely unrelated to sensory encoding is the creative generation of mnemonics as an aid in remembering random facts.

Man-made microprocessors employ encoding so that a small number of wires can be shared and made to send a large amount of information (for example, n wires carry 1 of 2^n binary codes). In a brain, however, binary codes are not generally advantageous, since there is an abundance of neurons for parallel connections.

The Amazing Neuron

A neuron can be schematically portrayed as in Fig. 1.5. Neural signals generally move from dendrites to soma, to axon, with many synaptic inputs, and many outputs of a given result at the tips of the axon. Although there is only one axon per neuron, there can be a large *fan-out*.

Boutons, Receptors, Spines

There are about 500 trillion synapses in the adult brain, and twice as many in a child. A synapse includes a presynaptic region, or a *bouton* that can release excitatory neurotransmitters from its vesicles as a result of an incoming pulse burst. Once released these are sensed by postsynaptic *receptors*, quite numerous, located mainly on *spines*. Synapses involve a narrow gap of about 20 nm between a

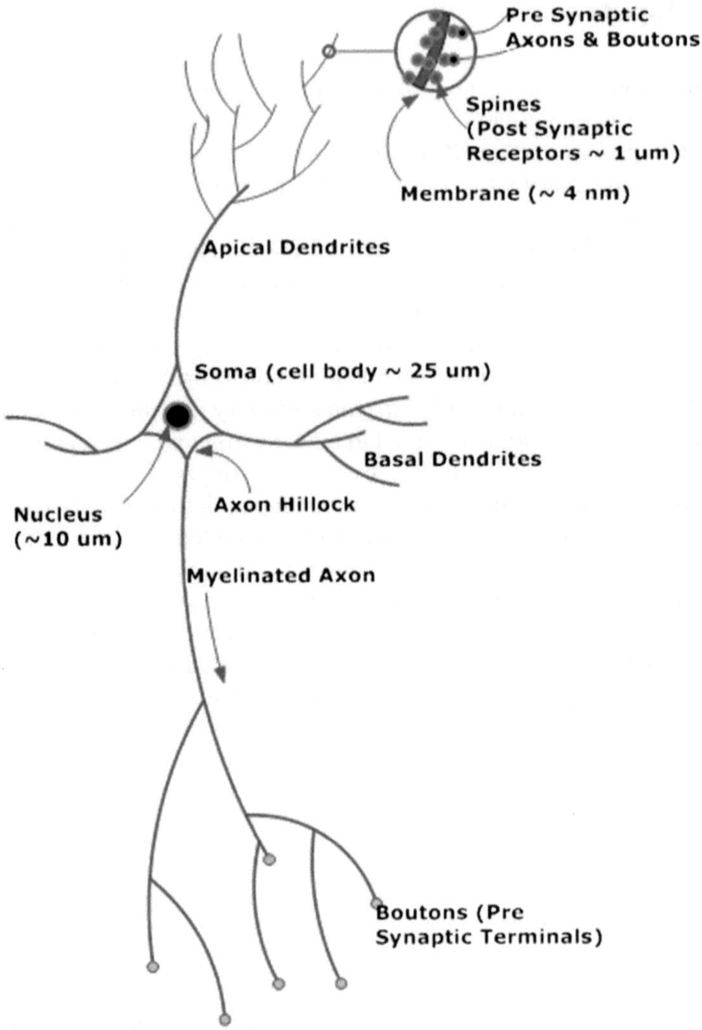

Fig. 1.5 Example pyramidal neuron (not drawn to scale)

bouton at the tip of a nearby neuron. Note that neurons never actually touch each other. Synapses typically transmit signals one-way, from bouton to receptor, in contrast to electrical conductors, which are bidirectional.

Dendrites, Soma, Axon Hillock, Axon

Pulses triggered in dendrites propagate according to the logic they undergo, traveling to the body or *soma*, where various incoming pulses may build up a charge. Once charge surpasses a threshold for triggering, pulses propagate from the soma and move along the axon; back propagations back down the dendrites are not unusual. The *axon hillock* serves to connect the soma to an axon in an efficient way, also minimizing interference from inconveniently located transmembrane protons. The *axon* is insulated from surrounding ionic solutions with myelination to facilitate a more rapid propagation of pulses along longer axons. Signals propagate down an axon but do not generally return by the same axon. When two-way communications are required, a separate neural axon is used for return signals.

Neural Signals

From a neuron's view, inputs are *excitatory neurotransmitters* directed toward *postsynaptic* receptors, and outputs are releases of neurotransmitters from *presynaptic* boutons at the tips or the terminations of axons. But from a circuits view, neurons are either activated or resting, and in this sense they are defined to be collections of binary devices, that is, complex gates for Boolean logic. A neuron is a biological cell with a membrane that accounts for its activity. When activated it produces a short burst of positive-going pulses [2, 3]. When resting, it holds a steady negative charge; that is, the interior is about −70 mV direct current relative to their exterior (direct current refers to a small but steady flow of charge leaking through the membrane).

A neuron may be quite large and is capable of arbitrary *Boolean logic*, which means arbitrary combinations of AND, OR, and NOT functions. The neuron supports differing styles of logic, such as dendritic logic and enabled (soma) logic, presented later in the book.

Neural communications are established via synapses; billions of complex neurons communicate together to achieve a system of consciousness. Solid-state logic is nothing like neural logic. Although also binary, solid state logic generally relies on DC voltage levels, a low voltage for false and a higher voltage for true, and does not depend on pulses, which makes it radically different from a biological system.

The output of an activated neuron is a low-energy, low-frequency *burst*, sometimes given the fuzzy term *action potential*. A neural burst consists of low-voltage pulses, below a few hundred hertz, charging and discharging membrane capacitance between about −70 and +40 mV; each pulse is roughly 2 ms wide. Neural signals take place within tens of milliseconds with typically ten pulses per burst, although this number can and has to vary in order to fit into a reasonably efficient system.

Conclusions

Even though this book is concerned mainly about logical structure, this chapter has reviewed the physical structure of a brain, at least the most obvious and interesting organs. Knowledge like this is important because it inspires conjectures about how parts make up a whole to achieve consciousness. For instance, a mini column in the cerebral cortex may be a candidate for an element of conscious short-term memory and its associated processing for a single-attribute; white matter may provide links to long-term memory and auxiliary processing as distributed in other organs of the brain, for instance, the amygdala and hippocampus. Accurate signaling paths are exceedingly complex and have yet to be established with certainty.

Medical pathology involving diseases of the brain has contributed significantly to what is known about brain organs. This chapter has mentioned specific diseases; common diseases as Alzheimer's and Parkinson's, thanks to those victimized, have greatly increased knowledge about the human brain.

References

1. Sylwester R (2005) How to explain a brain—an educator's handbook of brain terms and cognitive processes. Corwin Press, Thousand Oaks, CA
2. Arbib MA (2003) The handbook of brain theory and neural networks. MIT Press, Cambridge, MA
3. Byrne JH, Roberts JL (2009) From molecules to networks: an introduction to cellular and molecular neuroscience. Academic, San Diego, CA

Self-Study Exercises

(If you desire to keep a record of your efforts, please type your answers.)

Brain Structure

1. State briefly an expected function for each of the following parts of a brain, and also give a rough indication of location. (HINT: Refer to this chapter.)
   (a) Frontal Lobe
   (b) Broca's Area
   (c) Motor Cortex
   (d) Parietal Lobes
   (e) Occipital Lobes
   (f) Temporal Lobes
   (g) Somatosensory Cortex
   (h) Planum Temporale
   (i) Wernicke's Area
   (j) Cerebellum
   (k) Brainstem

(l) Hippocampus
(m) Substantia Nigra
(n) Thalamus
(o) Hypothalamus
(p) Olfactory Bulb
(q) Amygdala
(r) Cingulated Gyrus
(s) Corpus Callosum

2. Approximately how many of the following occur in an adult brain?

(a) Neurons (ANS. about 100 billion)
(b) Glial Cells (ANS. about 100 billion)
(c) Synapses (ANS. about 500 trillion in an adult brain)
(d) Sensory Inputs (ANS. One million in parallel)

3. For each of the following items relating to the forebrain, indicate briefly a location, and as appropriate, the structure, and a rough indication of the number of neurons involved.

(a) Cerebral Cortex (HINT: The chapter calculated about 25 billion neurons structured into various columns)
(b) Corpus Callosum (HINT: About 250 million axons connecting left and right brains)

4. State briefly what each of the following refers to. (HINT: Refer to this chapter.)

(a) Six Layers
(b) Mini Columns
(c) Macro Columns
(d) Brodmann Areas
(e) Glial Cells
(f) Gray Matter
(g) White Matter
(h) Distributed Model

Neurons

5. Explain briefly the purposes of each of the following parts of a neuron as presented in this chapter. (HINT: Refer to this chapter.)

(a) Boutons
(b) Axon
(c) Axon Hillock
(d) Soma
(e) Dendrite

(f) Spine
(g) Membrane

6. Plot a standard neural pulse burst according to what is given in this chapter. (HINT: A standard neural pulse will rise from about −70 mV to about +40 mV and then back to −70 mV, and will last about 2 ms.)

Chapter 2
Brain Architecture for an Intelligent Stream of Consciousness

Introduction

This chapter proposes a brain system building up from the known logical properties of neurons, which must fit together according to the principles of electrical system theory. The possibility of *reverse engineering* a human brain is not investigated below, since this sort of thing is generally attempted in a very limited way using lower life forms. The goal here is to explain higher intelligence and stream of consciousness.

It seems certain that cerebral neurons must operate together efficiently. This implies a system structure that would not be evident within the confines of molecular biology and biochemistry. System considerations soon lead to the concept of a distributed memory in which each element of memory is dedicated to a given mental attribute, an edge, shade, or color, for instance. For efficiency there has to be a one-to-one correspondence between short-term and long-term memory (LTM), since otherwise, memorization and recall are difficult to explain.

Basic psychology suggests that subconscious LTM is associative, where the recall depends chiefly on cues. Well known is that ambiguous cues lead to unpredictable recalls, while exact cues lead to exact recalls. Associative memory is a topic studied in computer science, the principles of which are likely to apply to human memory as well. Thus, as in any associative memory system, human memory must have valid cues for a memory search while avoiding mental blocks, and must resolve the problem of confusing multiple returns, by determining how much is permitted into conscious STM, and in what order.

Common sense is sometimes misleading when it comes to mental processes, but even so, the basic features of a mental system do not have to be exceedingly mysterious. For instance, *forgetting* can be viewed as a mental block in which cues fail to achieve a return from memory. Mental blocks are fatal in dangerous situations, so for survivability, it seems likely there must be a cue editor that fights mental blocks. The *cue editor* discussed in this book does this, and permits a

J.R. Burger, *Brain Theory From A Circuits And Systems Perspective*, Springer Series in Cognitive and Neural Systems 6, DOI 10.1007/978-1-4614-6412-9_2,
© Springer Science+Business Media New York 2013

memory search in the background for forgotten information such that forgotten things sometimes appear in one's consciousness minutes or hours later.

Mental confusion might result from multiple returns for a given set of cues, a common situation when an inadequate number of cues are used to call forth memories, and many different images are returned at once. For survival, and fundamental to any system of associative memory, human memory included, is multiple match resolution. First of all, it must prevent basic memory breakdown because of multiple returns on the same neural paths. Secondly, a basic *recall referee* is essential, including a processor that assigns priority to memory returns and then chooses the highest priority image to be gated into conscious STM. A basic recall referee is proposed in this book.

Decisions are important to a successful life form, and are shown to be choices made mainly by the brain, and not the person involved, based on past experiences logged in LTM. However, other types of decisions are only influenced by past memories, but not determined by past memories, such as random decisions. These are important when guessing a solution to a mysterious problem, for instance, as discussed below.

It continues to be a mystery how a person achieves inspired choices, such as common sense, truth judgment, intuition, and artistic appraisal. These seem to differ significantly from decision based logically on past memories, and decisions based on a random variable. In view of this mystery this book leaves open the possibility of subneural computing of some sort, possibly quantum computing, or quantum-like computing without an actual quantum system, since a quantum might be difficult to achieve biologically. Quantum computing, however unlikely it seems to skeptics, is too important to ignore completely, and so is introduced below and expanded in later chapters.

Classification of Neural Signals

There are many types of signals in neurons and synapses; the most important are illustrated in Fig. 2.1.

The *weak synapse* is one that results in a single pulse in the dendrite. Single pulses are needed in a practical system for accurate timing, as explained in subsequent chapters. For example, single pulses are needed in those situations in which *sequential* logic is necessary, for instance, when there are multiple returns from associative memory arriving sequentially that need to be parked in separate holding areas.

In contrast, a *regular synapse* produces a pulse burst of about ten pulses and is often used for *neural logic*. Neural logic here refers to neurons that compute Boolean *combinational* logic, equivalent to a network of AND, OR, and NOT gates. Included under logic are *buffers* and delay lines with a simple connective purpose.

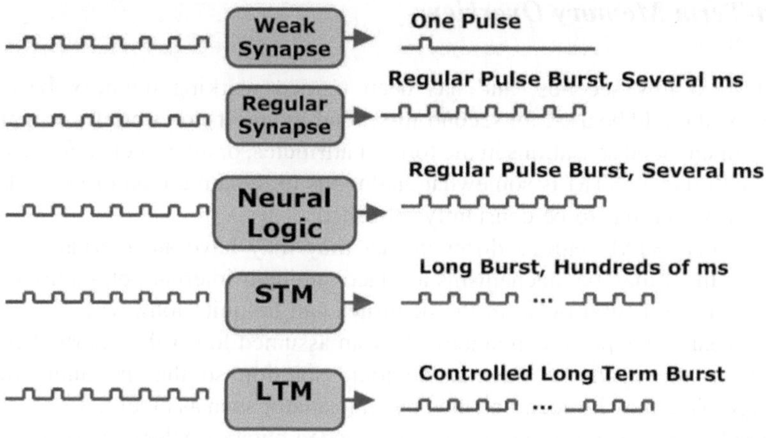

Fig. 2.1 Basic neural signaling (*ms* milliseconds, *STM* short-term memory, *LTM* long-term memory)

Inputs are thought of as pulses propagated along dendrites which have been triggered by receptors. Two major types of neural logic are identified in a later chapter: (1) dendritic logic, of which there may be thousands of gates per neuron, but in a gate of this type, pulses must arrive concurrently, and (2) enabled logic, which is limited to charge accumulation and subsequent triggering at a capacitive body such as a soma. Engineers are familiar with enabled logic in that it pertains to the activation functions used in artificial neural networks.

STM elements have an output burst that can be much longer than ten pulses, depending on ionic conditions within the dendrites, as discussed below. Bursts range in duration from a few milliseconds to several seconds. STM neurons produce regular pulses, that is, between about −70 and +40 mV, with pulse width about 2 ms, so that they are able to signal other neurons.

LTM elements ideally are instantly programmable, and last indefinitely, many decades. They may involve long-term potentiation (LTP) (trapped charge) and recursive neurons (that feed back on themselves); these may be organized to emit regular pulses only when called upon. Mechanisms and purposes of short- and long-term memory elements are going to be discussed in later chapters.

Introduction to Human Memory

The focus in this book is on explicit memory, as perceived in a brain, and not on implicit memory, such as automatic reflexes without direct mental involvement. Human explicit memory may be divided into two types, *conscious STM* and *subconscious LTM*. Signals that activate STM neurons are assumed in this book to flow from the five senses and also from recalls emanating from LTM.

Short-Term Memory Overview

STM lasts a few seconds, and has been termed working memory because if necessary, it could be used for second-to-second living. It stores briefly a significant number of encoded sensations in the form of attributes, or alternately a few complex organized images. STM is somewhat analogous to dynamic memory in a desktop computer, which has to be constantly refreshed.

Intuitively, STM neurons differ in that they may have an internal potassium deficit, or there may be mechanisms to inactivate their internal potassium ions [1]. Consequently a neural pulse in the dendrites can be quite long with an extended positive peak. This pulse is maintained by an assumed lower than normal conductance from the inside to the outside ionic solution so that it cannot quickly discharge. This may be accomplished by an insulator such as myelin. Consequently a long pulse excites the soma to emit an extended burst of pulses easily recognized by connecting neurons. With no extended burst the output is at rest, or false.

Long-Term Memory Overview

When conscious STM experiences images that cycle repeatedly, conditions may be triggered to permit them to be entered automatically into LTM. Images in LTM are not just visual pictures, but include a variety of sensory impressions, songs heard, words read, and impressions in general that have run through STM. What is entered automatically into LTM are particular collections of attributes that make up a mental image.

LTM is somewhat analogous to mass memory in a desktop computer in that it remains hidden until it is called upon. However, LTM in humans differs considerably from LTM in personal computers: Human memory is associative; all images are queried in parallel for matches to given cues. In contrast, computer memory generally depends on knowing a numerical address, and these are looked at serially, one at a time.

LTM and *learning* are two completely different things. Learning may involve synaptic potentiation and growth over a period of time; LTM can be formed immediately, too fast for growth, yet lasts practically forever.

For example, many of us have experienced dramatic or emotional events that occurred quickly, but are firmly remembered. This is a salient feature of LTM; they can be established practically immediately, and they can last indefinitely. LTM, as the term is used in this book, does not require synaptic growth. But it instead uses charge mechanisms that can react immediately. These mechanisms seem to be very different from those of STM neurons.

The exact mechanisms for LTM are still being discussed. How they work is not obvious because neurons do not release many signals compared to solid-state devices, making them difficult to observe in action, especially since they probably are active for only a brief time, and then only if triggered. Generally speaking, neurons are relatively small and easily upset by probing, so direct in vivo observation is difficult. Note for the record that memory elements are located randomly and duplicated extensively, so given memory episodes are nearly impossible to locate experimentally, meaning particular unwanted memories are difficult to remove surgically.

An interesting theory is that needed attributes are held efficiently by *LTP* [2]. Another interesting theory is that recursive neurons support cycling pulses that use almost no energy, and that are readable with positive signals for needed attributes [3]. A hybrid element using LTP and recursive neurons is proposed in a later chapter, giving instant memorization as well as efficient long-term storage. Elements of LTM as needed for a neural system have to be readily accessible to conscious STM, and it is critically important that they have ability to be queried; so a logical system is suggested.

Introduction to a Memory-Based Brain System

When considering a neural system that results in a stream of consciousness, it may be noted that nature evolves according to certain principles. Entities of nature, including brains, can survive only if, on the average, they are efficient.[1] Concepts such as the *grandmother neuron*, in which all attributes are stored in a single large neuron, have to be discounted because they are incompatible with bringing forth an image into conscious STM in an efficient way.

Transferring images from LTM into STM, and transferring images from STM into LTM must also be efficient. This implies that each attribute located in STM directly corresponds to a similar attribute located in LTM, an architectural feature known as *distributed memory* [4] to be discussed later.

This one-to-one correspondence implies that memory is logically organized into words. Memory words are a logical, not a physical, structure. Since not every image contains every possible attribute, memory words are *sparse*. Sparse means that entries in a word are separated by unused places corresponding to where certain attributes are not present, and that these places hold either unused neurons or neurons that have been turned to some other purpose.

Excitatory synapses serve to initiate neural pulses; inhibitory synapses serve to moderate neural pulses, as does the capacitance of larger bodies such as the soma, all of which results in a pulse-based system of logic. Memorization and other brain capabilities depend on *dendritic logic* which occurs mainly in dendrites where *Boolean operations* are accomplished, as verified in part by simulations of dendrites.

[1] Maupertuis principle of least action that nature is thrifty in all of its actions.

Another form of logic, known as enabled logic, occurs for larger bodies of capacitance, for instance, somas, as presented in a later chapter. As a result, a complete set of Boolean logic (AND, OR, NOT) is available, a natural result of synaptic activity in combination with active and passive regions in dendrites, soma, and sometimes also axons. Dendritic logic is compared to enabled logic in a later chapter.

Systems are typically expressed in terms of logical, not physical, structures, and brain systems are no exception. Human LTM is *associative*, which hints that location is not a prime consideration; things are recalled not according to where they are, but according to what they are. In contrast, in a desktop computer, each word in memory has a unique *address*, and a physical place. Addresses are used for reading or writing to locations that may be chosen at random, so the choice is random. Hence the term *random access* memory, or RAM. Human memory, in contrast, does not have fixed addresses, so we know a lot less about where memories reside.

Retrieval from subconscious LTM begins with cues. For example, seeing someone's dog can bring forth a memory of your own faithful companion, and might even bring forth details of face, color, size, and so on. Other such examples are visiting your home neighborhood which brings forth memories of stores, houses, rooms, passageways, and so on; seeing a book that you used in college can bring forth memories of what you learned in those days. Cues are essential to memory retrieval.

Cues need to be chosen judiciously from the current contents of conscious STM. If the available cues do not result in a recall, editing is available to adjust the cues for another try. Hopefully there will be a *hit*, which means an exact match to the cues. Memories with attributes that exactly match the applied cues will be read for possible entry into conscious STM. Cues applied to LTM have to be exact, since LTM are called forth only for exact matches. Gratefully, old memories are dormant until receiving an exact set of cues. Serious overload would result if we had to experience all facts, even irrelevant ones, at all times.

Under the distributed model each attribute connects to specific locations in STM and corresponding locations within words of LTM. These locations exist physically, although current technology does not permit a knowledge of exact location. But nevertheless each location in STM has corresponding locations in LTM that are reserved for a particular feature. This is an efficient way to organize STM and LTM because it avoids extra encoding and the passing of data to intermediary storage sites.

A given attribute might be said to have meaning not because it has any special coding, but because of its physical location in a word. For example, the color *red* might be in a certain location, say the 537th cell from the left in a delineation of logical STM; it is also the 537th cell from the left in each of its replications in LTM. Location in memory and the context in which the attribute is used contribute to realizing the meaning of an attribute, as for example, when seeing a fire engine in the context of a street, it is expected to be red.

The one-on-one distributed model implies that neurons are not multiplexed, that is, different attributes are not communicated by the same neuron. Fortunately neurons are quite plentiful and are conveniently dedicated to a given attribute as described.

Human memory is not *erasable* and is not converted into *blank* words in the ordinary sense, nor can words be overwritten as they are in the *RAM* of desktop computers. Once something is committed to memory, it is there indefinitely. In this sense, LTM is akin to *read only memory* (ROM). Under the ROM concept, new brain memories cannot overwrite old memories; instead they must be written into blank words available for that purpose, the subject of a later chapter.

A System for Stream of Consciousness

A system of consciousness can now be sketched: Simulations of neural logic have suggested a brain system based on *bottom-up* construction. *Attributes* are calculated from the senses and are made to enter conscious STM as an image, but alternated with the senses are images in the form of a set of attributes from memory recalls. Attributes currently in conscious STM are a main source of cues, which are not necessarily perfect.

For *conflicting* or *overdetermined* sets of cues, or too many cues, editing is essential. Editing strives to remove conflicting cues to achieve returns. For *underdetermined* cues, when cues are few in number, one expects an excessive number of returns. In this case, returns are edited subliminally, and the one that calculates to be most important is selected for entry into STM, where it becomes a *recall*. Figure 2.2 shows how these two major editors form what may be construed to be a cybernetic[2] system, that is, a system in which attributes selected from a current image are used to determine the next image.

A system for stream of consciousness may be summarized as follows. A set of attributes in STM are taken as cues and serve to find related images in associative LTM. A high-priority recall is selected from the resulting returns, and overwrites the fading contents of STM. The process repeats, giving a moving picture in conscious STM.

Cue Editor

What can happen is either (1) no recall or (2) excessive number of recalls. Everyone has experienced no recall, trying to remember something but being unable to do so. So you go about your business, but unknown to you, a search proceeds subliminally for what you are trying to remember. Cues keep cycling subconsciously, but are insufficient to return an image. However, when you least expect it, possibly at an inconvenient moment, the correct memory will pop into conscious STM with amazing clarity. This is an indication that the brain has worked in the background

[2] Cybernetics was defined by Norbert Wiener (1894–1964), in his book of that title, as the study of control and communication in the animal and the machine.

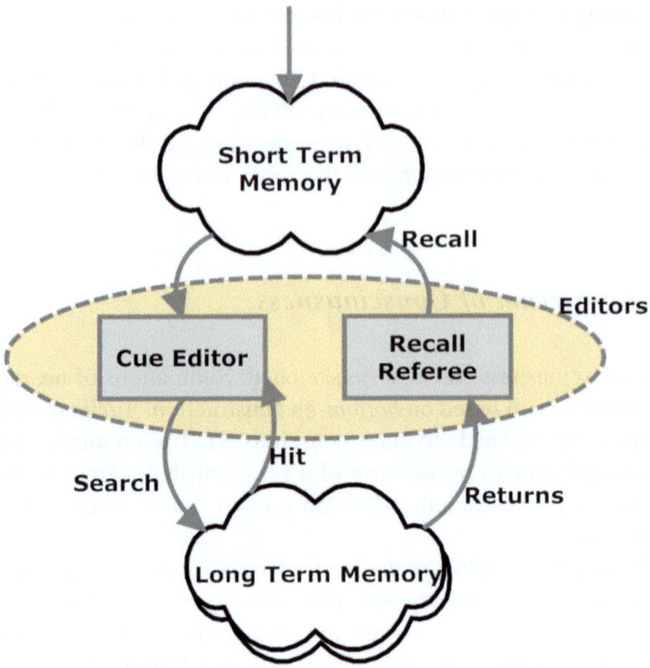

Fig. 2.2 Subliminal editors in relation to short- and long-term memories

without being noticed, adding or subtracting cues from a search until the cues are exactly right, as they must be for a recall.

A *cue editor* helps resolve conflicting cues (which would result in no matches in long-term associative memory). The term *hit* in the above figure means that a return has occurred. But if *hit* is false, it is the task of the cue editor to modify the cues slightly so as to produce a return. As new images from the senses appear in STM, new rounds of cue editing are necessary. The past unfinished editing process moves into the background but keeps working. This is why a person sometimes belatedly recalls a forgotten piece of information. A subsequent chapter goes into circuitry for a cue editor.

A cue editor is not in a position to judge the priority of multiple returns. This is the task of a *recall referee*.

Recall Referee

An underdetermined set of cues generally results in a plethora of returns (think of everything you know that contains the color green, for instance). This overabundance has to be processed to assess the priority of each. If something is critical, it must be passed quickly into STM, because survival may be involved. For instance, safe responses in the event of danger must be promptly directed into conscious STM.

A technical term for what the recall referee must accomplish is *multiple match resolution*. Multiple match resolution avoids neural overload while assessing each return, calculating a priority value for each. Priorities are computed in parallel to speed the process of recall. The highest priority images are gated into conscious STM, giving a stream of consciousness one image at a time.

Returns may occur many per second, so a lot of searching is possible subconsciously. They come forth in rapid succession and are parked in biological registers where calculations are accomplished. The priority computation will contain emotional factors, and survival factors, and others integrated into each word of LTM [5, 6]; these have been well discussed [7]. Priorities are compared. That image with the highest priority is gated into conscious STM to become one's next thought.

Recall editing is accomplished with a novel method of parallel processing using recursive neurons operating as controlled toggles. All this is discussed in subsequent chapters.

The Nature of Decisions

Decisions Based on Past Experience

There is a theory that decisions are made not by *free will* in the sense of personal spur-of-the-moment choices, but by a search process that depends on past experiences and training. Often a ready-made decision is available based on past similar situations held in LTM, as for example when you are asked to donate to a cause that you do not like. A "no" answer is preprogrammed. Decisions, therefore, are affected by past experiences, what you did in similar situations, and whether or not an action supports your basic philosophy.

There is some evidence that free will is an illusion [8, 9]. Electroencephalography concerning the timing of finger movements, published by Benjamin Libit and others, indicates that choices are made in the brain without knowledge by the person involved; decisions were made well before a person realized it. Surprisingly, the brain seems to be in control, making decisions, and eventually informing consciousness. It is not the other way around, in which a "person" makes a conscious decision (without the aid of his or her brain), and then informs the brain.

Decisions with a Random Variable

It may be noted that a brain appears to search itself continuously in the background not only for forgotten facts and situations but also for solutions to problems. A problem could mean a real problem with no easy logical solution, for example, trying to open a complex combinational lock without knowing the combination. Trying all possibilities systematically requires excessive time. It is more practical to try random combinations and trust to luck (and intuition). The brain may very well solve real problems this way, by random searches for quick solutions.

Random variables come into play in a *cue editor*. Simple circuits can randomly remove cues, search memory, and restore them while randomly removing others to obtain a return. Having good cues is very important. A cue editor is a logical subcircuit that can store a basic set of cues, if they are conflicting, and edit them in a random way, in the background.

As further indication that memory searches occur randomly, dreams, as everyone knows, are brief illogical episodes usually soon forgotten. A person is dimly aware of dreams during light sleep partly because there are no overpowering sensory images to block them out, as there would be during the day. Dreams, according to one theory, are a result of the repeated application of random underdetermined cues and multiple returns, some of which pass through the recall referee into STM because of their emotional content. Repetition of dreams sometimes causes them to be noticed as a person wakes.

Brainstorming may be like this, in which a random attribute is applied to a cue editor that causes unexpected returns. There usually are many returns, but only the most recent, most alarming, or most interesting scenarios are permitted into conscious STM by a recall referee.

Inspired Decisions

Currently there is no way to explain, let alone design machines that have common sense, truth judgment, understanding, artistic appraisal, and other hallmarks of human intelligence [10]. There seems to be a hidden source of computational power. Visionary pioneers have surveyed the possibility of quantum computing within a brain [11]. No doubt there are important quantum mechanical behaviors within ion channels and within synapses and elsewhere, since ions and their electrons are small and subject to quantum theory. One interesting hypothesis suggests quantum computing within the microtubules of neurons, proposed to generate higher consciousness [12, 13]. Others see quantum tunneling of electrons between synapses as creating consciousness [14]. For anything new, there are skeptics [11, 15, 16].

Quantum computations in neurons are an unproven hypothesis. Nevertheless quantum mechanics is a valuable metaphor for certain brain behaviors [11, 12]. Thinking about quantum mechanics is fun, and exercises the brain, which is a healthy activity. A later chapter discusses recent avenues in the search for brain quantum computing.

In contrast to the qubits of quantum theory, it may be possible for recursive neurons to take on qubit properties, such as holding true and false simultaneously with given probabilities in a probability space. Recursive neurons may also serve as controlled toggles for parallel processing, such as needed in a recall referee. Recursive neurons configured to act like qubits are termed *simulated qubits*. They may be a source of enhanced computational power, particularly if formed at the molecular level within a neuron. Such a discovery might actually be more likely than real qubits, which require an isolated quantum system.

Conclusions

This chapter has presented a model in which the various attributes of a mental image each have their own fixed locations. Also each attribute has one or more backup locations, but physical locations are irrelevant in the logical brain architecture of this book.

Conscious STM is modeled as fed with appropriately processed signals from the five senses, interspersed with successful recalls from subconscious LTM. A person is aware of mental images mainly from the senses, but interspersed are recalls from LTM that are extremely important to survival and intelligent behavior.

A cue editor helps ensure the availability of recalls when cues are inadequate. This book outlines a model in which a cue editor works with selected attributes from a current image in conscious STM. When there is a mental block, the cue editor immediately begins to modify cues in a random way to obtain returns from LTM. A cue editor serves to resolve conflicting (overdetermined) cues, and cues for which there are absolutely no memory matches within associative LTM.

Under this model, ineffective or conflicting cues are not discarded but are stored in a special register composed of recursive neurons. Before searching again, a few cues are randomly removed, and the remaining are used for a new search; and if still there are no matches, then the removed cues are restored. The process of randomizing is repeated again and again until after some point, returns are finally obtained. This system can account for the common experience that forgotten information often pops into one's head with delay, perhaps much later, at an unexpected moment.

Parallel to the neural paths for memory search are neural paths that bring up returns from LTM. A recall referee serves to deal with the common problem of multiple returns from a given set of cues. Multiple returns are expected to occur in a group shortly after cues are applied. Within a recall referee, each return undergoes a quick but exact digital determination of priority. Most returns are rejected as being unimportant, but one will have maximum priority and will be gated directly into conscious STM, where it results in physical actions or additional memory searches.

Methods for priority calculations are presented in later chapters using recursive neurons operating as controlled toggle devices, all operating in parallel. Controlled parallel toggles calculate priority for each multiple return in a prompt manner.

Brain editors as above operate between short- and long-term memories to make decisions and to produce a stream of consciousness, a stream that is infinitely more intelligent than mere dumb acceptance of bad cues and unimportant returns. Subsequent chapters are going to develop a structure with appropriate circuits for short- and long-term memory as well as neural circuits for editing to accomplish the abovementioned system.

References

1. Kandel ER (2006) In search of memory: the emergence of a new science of mind. W. W Norton & Co, New York
2. Bliss T, Collingridge GL, Morris RGM (2004) Long-term potentiation: enhancing neuroscience for 30 years. Oxford University Press, Oxford
3. Burger JR (2009) Human memory modeled with standard analog and digital circuits: inspiration for man-made computers. Wiley, Hoboken, NJ
4. Kanerva P (1988) Sparse distributed memory. MIT Press, Cambridge, MA
5. Franklin S (1995) Artificial minds. MIT Press, Cambridge, MA
6. Anderson J (1983) The architecture of cognition. Harvard University Press, Cambridge, MA
7. Dennett DC (1991) Consciousness explained. Back Bay Books, Boston
8. Wegner DM (2002) The illusion of conscious will. Bradford Books, Cambridge, MA
9. Restak R (2006) The naked brain: how the emerging neurosociety is changing how we live, work, and love. Harmony Books, New York
10. Penrose R (1989) The emperor's new mind: concerning computers, minds and the laws of physics. Oxford University Press, Oxford
11. Tarlaci S (2010) A historical view of the relation between quantum mechanics and the brain: a neuroquantologic perspective. NeuroQuantology 8:120–136
12. Hameroff S, Penrose R (2003) Conscious events as orchestrated space-time selections. NeuroQuantology 1:10–35
13. Hameroff S (2007) Orchestrated reduction of quantum coherence in brain microtubules: a model for consciousness. NeuroQuantology 5:1–8
14. Walker EH (2000) The physics of consciousness: the quantum mind and the meaning of life. Perseus, Cambridge, MA
15. Tegmark M (2000) The importance of quantum decoherence in brain process. Phys Rev E 61:4194
16. Donald MJ (2001) Book review, Walker, the physics of consciousness: the quantum mind and the meaning of life. Psyche, vol. 7, www.bss.phy.cam.ac.uk/~mjd1014, accessed March 4, 2013

Self-Study Exercises

(If you desire to keep a record of your efforts, please type your answers.)

1. Sketch on graph paper the four major types of bursts found in neurons according to this chapter. (HINT: Single pulses, bursts of about ten pulses for basic communications and logic, extended bursts for up to a few seconds for STM, and indefinitely long bursts, or bursts that can be triggered and stopped if and only if there is an attribute of LTM)
2. List the characteristics of and possible mechanisms for:
   (a) STM (HINT: Lasts no more than a few seconds; could be caused by a shortfall of available internal ions thus resulting in long dendritic pulses; this would trigger long pulse bursts of regular pulses in the soma)
   (b) LTM (HINT: Formed instantly and lasts forever; could be based on quick LTP, or recursive neurons that cause a low-profile pulse to cycle indefinitely, or a hybrid of potentiation and recursion)

3. Explain briefly why the following components are important to humans.

   STM (HINT: For second-to-second living and consciousness)
   LTM (HINT: For knowledge, intelligence, learning from past experiences)
   Cue editor (HINT: For survivability, to guard against memory blocks)
   Recall referee (HINT: For survivability, to avoid confusion and focus only on high-priority returns from LTM)

4. Explain what a cue editor does? (HINT: A cue editor keeps track of cues that fail to create a return, and changes them a few at a time in an attempt to make them work.)
5. Explain what a recall referee does? (HIINT: A recall referee computes a priority value for each image returned for a given set of cues, and permits only the highest priority image into conscious STM.)
6. What is meant by a memory-based system? (HINT: All possible responses and knowledge relate to LTM.)
7. Describe three categories of brain decision making. (HINT: Past experiences, random choices, inspired choices)

Chapter 3
Circuit Elements Required for Neural Systems

Introduction

Circuit elements introduced in this chapter are (1) short-term memory (STM) neuron, (2) regular synaptic contact, (3) weak synaptic contact, (4) postsynaptic capacitor with initial charge representing long-term potentiation (LTP), (5) capacitive termination for dendrites, (6) dendritic AND, OR logic, (7) dendritic XOR, NOT logic, (8) enabled, or soma AND, OR logic, and (9) enabled, or soma NOT logic. Also available are delay and sundry other elements. Some of these are novel in traditional circuits and systems, which has always been aimed at solid-state circuitry, but these nine elements are essential when modeling pulsating neurons. The functions of these nine elements are demonstrated by a set of simulation experiments in the appendix.

Classical elements are also useful, such as capacitance, resistance, independent voltage sources and current sources, and controlled sources controlled by circuit *variables* (voltages and currents). Circuit elements as above are required to (1) serve to model neural behavior in an accurate way and (2) satisfy the needs of a functioning brain system.

Models as in this chapter summarize experimental data; they serve to inspire future experiments, and hopefully encourage understanding. Models, of course, must be routinely updated to reflect new experimental data. Models sometimes lead to practical versions of the modeled system, as they have done for artificial neural networks, but students are well advised to seek full comprehension before attempting engineering applications.

J.R. Burger, *Brain Theory From A Circuits And Systems Perspective*, Springer Series in Cognitive and Neural Systems 6, DOI 10.1007/978-1-4614-6412-9_3,
© Springer Science+Business Media New York 2013

Introducing…The Active Neuron

The neuron or at least a model of one will now be introduced. The brain contains a great many neurons, estimated to be 100 billion more or less (1 billion $= 10^9$). With this many, it is only natural that some neurons will differ from others not only in shape but also in makeup and function.

What is known about neurons follows from a combination of measurements in the laboratory and from computerized simulations using realistic physical parameters. It is found that neurons can be triggered to give pulses between about −70 and +40 mV; each pulse is roughly 2 ms wide; these propagate back and forth along dendrites, and down the axon to boutons, also known as terminations or tips. Subsequently in the presynaptic region of a synapse, these boutons release neurotransmitters into the synaptic cleft. Neurotransmitters, like any larger molecule, can be ionized, which makes them more effective in making their presence felt. They strongly impact a postsynaptic region.

In order to activate a neural burst, there needs to be a force that triggers molecules in the membrane. Excitatory neurotransmitter ions are one way to accomplish triggering in a receptor, usually on a spine located on a (basal or apical) dendrite. These in turn connect to the soma (or body) of a neuron. Note that spines and synapses are very numerous but not all are necessarily employed at the same time for neural operations.

Pulses may also be triggered by internal potentials. The membrane of the soma itself, as well as that of the dendrites have an electrical threshold for activation. Once the charge across the membrane capacitance exceeds this threshold, triggering occurs. Pulses continue to be produced as long as the membrane voltage is above a triggering threshold, which amounts to about 15 mV above a rest level at −70 mV, with triggering at roughly −55 mV.

A membrane becomes active because it is bounded inside and out by ionic solutions. It can be triggered by internal voltage to generate pulses across the capacitance of the membrane, pulses that have a characteristic waveform (shape, amplitude, and width). Once a pulse forms, it triggers adjacent areas of a nonlinear sensitive membrane and appears to propagate without attenuation, albeit much slower than a pulse propagates in linear media.

Technically each pulse propagating along exposed dendritic membrane is a *soliton*. By definition, a soliton is an electrical pulse that propagates and is reenforced as every step of the way by nonlinear media to maintain a given shape even though there are significant losses in the conductor and the membrane. Without this important regeneration effect, pulses in dendrites and also axons would soon die out, and brains as we know them would be impossible.

Membranes can be deactivated and rendered passive by cutting off their access to ions. Important categories of Boolean logic in dendrites depend on regions of active membrane alternated with regions of passive membrane. For example, an AND gate results when one signal cannot get through a passive region, but two signals applied simultaneously can. That is, two signals merging at a branch can combine to charge a passive region and thus exceed a threshold for triggering in a region that lies just beyond the passive region.

The axons of a neuron, in contrast to dendrites, have membranes that are chiefly passive because they are covered with myelination, or white matter that insulates them from the surrounding ions. With membranes blocked, axons are linear conductors with distributed series resistance, and reduced membrane shunt conductance and capacitance. So losses are lower. The speed of pulses is higher.

Signals move faster in axons since capacitance is lower. Eventually there is some loss of energy because of axon series resistance, but pulses in neurons are reenforced every millimeter or so by the exposed regions at the *nodes of Ranvier*. Each node consists of 1 or 2 μm of exposed membrane that serves to restore attenuated pulses.

What Membrane's Do

Neurons are enclosed in a membrane that in turn is surrounded inside and out by thermally active ionic solutions. Generally one may imagine sodium ions (Na^+) on the outside, and potassium ions (K^+) on the inside, although other elements are possible. All particles, ions included, undergo chaotic thermal activity; they vigorously speed about, bouncing off each other and also neural membranes. This chaotic activity represents thermal energy and relates to temperature for any temperature above absolute zero (−274°C).

Consider a membrane in equilibrium with given type of ion, sodium, for instance, such that there are different concentrations on one side compared to the other. Assume ionic concentration (ions/cm^3) on the outside much higher than on the inside. As sodium ions collide with the outside of the membrane, they result in a transfer of electrically positive charge to the inside.

Positive ions have been visualized as penetrating into the interior by diffusion through the membrane, or through ion channels, building up positive charge, and ion population inside. Equally possible is that electrons from the interior are tunneling into the membrane, and into ion channels, and being captured by external Na^+ ions, so as to leave behind a positive charge. If there is a positive voltage built up inside, lower speed positive ions from the outside will be reversed and may block channels, thus terminating charge transfer. Overall the end result is a certain voltage positive on the inside relative to the outside. When all ions types are considered there will be a certain *resting voltage* inside relative to the outside.

Potassium ions on the inside have their own effect, which turns out to be a larger effect than sodium, resulting in negative voltage on the inside. When all ions are considered, the net resting voltage is about −70 mV inside, relative to outside of a typical neuron. And since a membrane is conductive, there is current and energy dissipation that has to be offset by a process of "metabolism".

Since membranes are thin, perhaps as thin as 5 nm (5×10^{-9} m), there is an electric field of roughly 140 kV/cm. This is roughly five times what it takes to cause lightening in air (about 30 kV/cm breaks down air). So electrical stress is significant. All molecules in a membrane are polarized to some extent and held by electrical force in an elongated fixed position until triggered.

Pulses Brokered by a Membrane

To begin, the interior voltage of a neuron is at rest, −70 mV. A neural membrane consists of a phospholipid bilayer of molecules that is profusely and randomly textured with larger transmembrane proteins (they number about 10^{10}/cm^2 roughly). One way to trigger a membrane is to impose a modest reduction in the voltage across the membrane, from −70 to −55 mV. This will begin to unlock some of the larger transmembrane proteins in the membrane and to open channels of conduction. Thus begins a neural pulse.

Sodium ions from outside the neuron penetrate into the unlocked membrane, capturing electrons, and resulting in positive charge accumulation inside. Particles have been estimated to go into and through a membrane at about 10^{15} particles/s/cm^2 [1]. This converts to amperes by multiplying by the charge of one electron, 1.6×10^{-19} C, Coulomb/electron, and then multiplying by membrane area in square centimeters. The resulting current charges membrane capacitance in a positive direction.

Voltage is determined by a basic formula: $\Delta V = I\Delta t/C$, where ΔV is the increase in voltage in millivolts, I is charging current in microamperes, C is membrane capacitance in microfarads, and Δt is the amount of time used in milliseconds. Voltage build up across a membrane depends largely on the duration of the current pulse. Charging current has been observed to be fairly steady, as evidenced by photographs of neural pulses, and also computerized simulations, more or less like the curve in Fig. 3.1, whose rising voltage increases at a constant rate of about 140 mV/ms.[1]

Note that once triggered, the pulse goes smoothly up to its peak and then smoothly down to its undershoot value. Subsequently it slowly drifts back to a resting potential. All occurs within a few milliseconds; this simulation indicates a risetime of about 1 ms; a slightly longer falltime, perhaps 1.5 ms; and a longer recovery time of perhaps 2 ms. This recovery time and other variables define a *refractory* period during which a neuron, if triggered, will not give a full output.

Relevant to explaining a neural pulse is that charging current is maintained until internal voltage reaches about +40 mV. At this point the electric field applied to the molecules in the membrane is strongly reversed. The molecules now change orientation so that the sodium current begins to cut off. Once sodium charging is

[1] Note that slower positive ions in a channel would be stopped and reversed by the repelling internal positive electric field, blocking the channel. But experimentally the average rise in current is steady. This indicates capturing of electrons from the interior, allowing the pulse to maintain a steady rise.

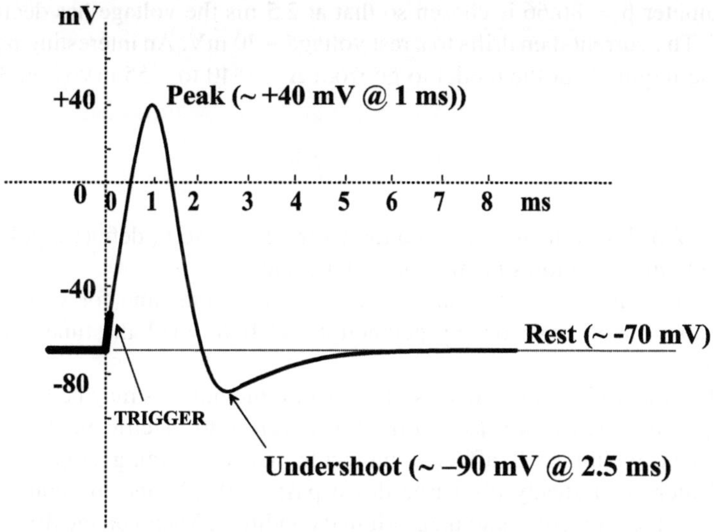

Fig. 3.1 Reference pulse, internal voltage relative to outside

cut off, neglecting loss conductance and neglecting the effects of potassium and other ions, internal voltage would hold at a steady value.

Potassium is also active. The larger potassium ions inside the neuron have their own effect, moving positive charge away from the interior. This occurs at a lower rate of charge transfer because potassium ions are larger and move slower at a given temperature. Once the sodium charging cuts off, the potassium are dominant, effectively discharging membrane capacitance to below the rest voltage.

Photographs indicate an undershoot of a few tens of millivolts below −70 mV. It is conjectured that this is necessary in order to snap the molecules within the membrane into their original locked position and thus put an end to the given pulse.

A Simple Pulse Model

A simple model is useful as an aid to understanding a neural pulse. The charging current is

$$v(t) = -70\,\mathrm{mV} + \alpha t \qquad t \le 1\,\mathrm{ms}. \tag{3.1}$$

Time (t) is in milliseconds and $\alpha = 110$. Thus at 1 ms the voltage is +40 mV. After a millisecond the discharging current is modeled as equal to

$$v(t) = +40\,\mathrm{mV} - \beta(t-1) \qquad 1\,\mathrm{ms} \le 2.5\,\mathrm{ms}. \tag{3.2}$$

The parameter β = 86.66 is chosen so that at 2.5 ms the voltage has decreased to −90 mV. The current then drifts to a rest voltage −70 mV. An interesting parameter is the time required for the model to go from Δ = +40 to −55 mV (Fig. 3.2):

$$t_\Delta = 1 + \frac{95}{\beta} \text{ ms} \tag{3.3}$$

At −55 mV or less a membrane is no longer triggered, so t_Δ defines a pulse width. This simple model ignores membrane conductance.

Note for future reference that as beta decreases, the time for which voltage remains above −55 mV goes up geometrically! If $\beta = 0.1$ the time involved is about 1 s.

Another way to look at a pulse is that a neural membranes may be thought of as following a hysteresis loop as in Fig. 3.3. *E* refers to electric field and *D* to a displacement within each affected molecule. Once charging stops at roughly +40 mV there is a steady discharge down past −70 mV and to about −90 mV which snaps the molecules into their original condition. Then voltage drifts to a rest level, about −70 mV. The hysteresis loop is a common property of materials and portrays membranes from a different point of view.

Delay Elements

Delay in dendrites is very roughly 15 ms/cm. Signal delay is determined by the length of the path and also the membrane capacitance of the path, and such current-charging parameters as the density of conductive pores (or ion channels) in an unmyelinated neural conductor and also by local ionic concentrations. Inhibitory

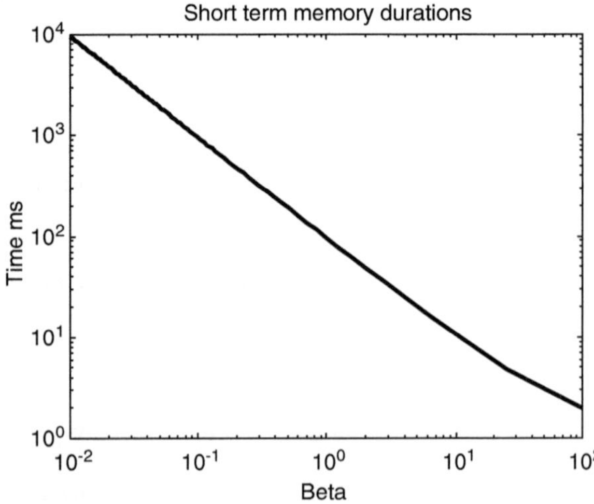

Fig. 3.2 Time (ms) to move from +40 to −55 mV versus β

Fig. 3.3 Ferroelectric (hysteric) characteristic (not drawn to an accurate scale)

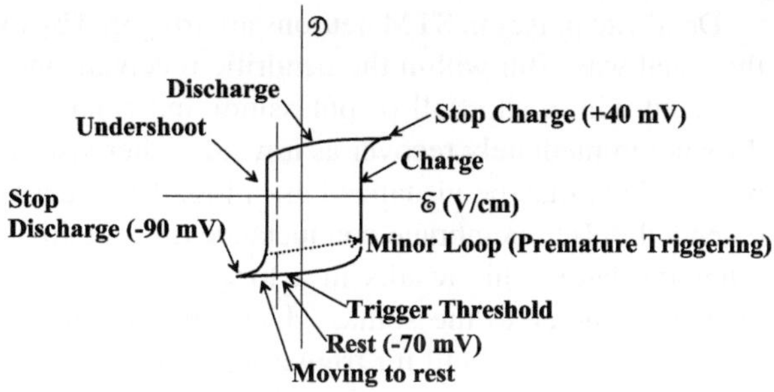

Fig. 3.4 Delay and short-term memory symbols

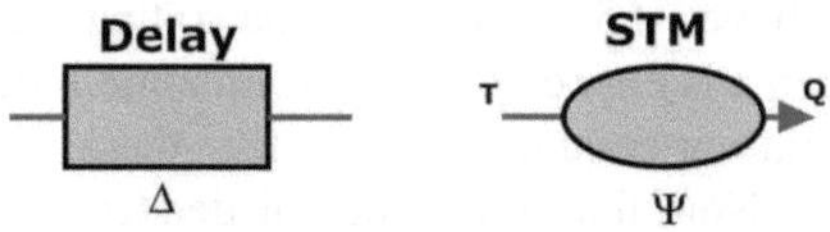

neurotransmitters also affect delay. Delay is denoted by a simple rectangle (Fig. 3.4) with an input and an output and with the word *Delay*. Signals in dendrites could theoretically go either way, left-to-right or right-to-left. In a neuron, signals tend to flow from a synaptic input to a bouton output, in which case an arrowhead is sometimes used to indicate the direction of signaling. If relevant or critical, the amount of delay is specified to be a value represented by Δ.

Short-Term Memory Neurons

Neurons often emit perhaps ten pulses per burst lasting a total of a few milliseconds. In contrast, STM neurons are modeled as producing bursts extending up to several seconds. This is because a long neural burst may represent a STM for a given attribute, such as shape or color. A selective collection of such features, of course, builds an entire image.

STM neurons are assigned the oval symbol in Fig. 3.4. The *duration* of a STM burst is assumed variable, ψ, from milliseconds to seconds. This neuron is triggered at T and produces a pulse burst at Q; the direction of the signal is usually understood from the context in which it is used. An STM neuron is generally triggered via a synapse, as opposed to a conductive contact, because STM dendrites are unique; they support longer pulses and so must have lower membrane loss. Note that both inputs and outputs are bursts of pulses; a short trigger to the input T and usually a longer burst from the output Q.

In addition to representing memory attributes, STM neurons serve to produce timing signals in a neural system. The duration ψ is fixed at some biologically useful value that does not change easily, since it depends on dendritic parameters.

Dendritic pulses in STM neurons are triggered by excitatory neurotransmitters in the usual way. But within the dendritic receivers and the dendrites themselves are modeled with a shortfall of potassium and calcium, so that the dendritic voltage does not immediately recover as it would otherwise. This results in a long dendritic pulse, which may be visualized as in Fig. 3.5 as derived from Fig. 3.2. To have an extended pulse, membrane conductance must be low, insulated by myelin or some other insulator. This results in a longer lasting internal charge that becomes an extended trigger to the soma, which pulsates until the dendritic pulse's voltage drops below a threshold for triggering.

Holding charge to a soma produces what has been termed a *maximum rate* of pulsing. Maximum rate of pulsing occurs by applying a steady triggering voltage above −55 mV to the soma so that it is continually retriggered as soon as possible. This is expected to result in some reduction in the amplitude of the output pulses, but not enough loss to render them ineffective in signaling other neurons.

Note that a shortfall of dendritic receiver potassium and calcium implies that electrical current is unavailable to reduce in a timely way the dendritic pulse after it reaches and holds at +40 mV. But over time, a trickle of charge will penetrate through the membrane and reduce internal positive voltage as electrons return to the interior. Thus STM fades in a controlled way.

Modeling as a shortfall of internal potassium is probably a simplification. Experiments concerning implicit memory (reflex memory) with a sea snail called Aplysia were conducted over a period of several years by Kandel [2] and others. It was found that an interneuron can serve to release serotonin near the synapses of an ordinary neuron, giving a memory effect.

Theory is, receptors on a neural membrane receive serotonin and convert it into substance known as *cyclic AMP* inside a neuron. Cyclic AMP is what results in a measurable memory effect. Cyclic AMP inside a neuron shuts down potassium channels, with the end result there is an increase in the period and quantity of glutamate (excitatory neurotransmitters) released from the neuron. Releases can last from milliseconds to minutes. This work was about implicit memory, whereas brain STM is explicit memory, but it may somehow apply to STM neurons.

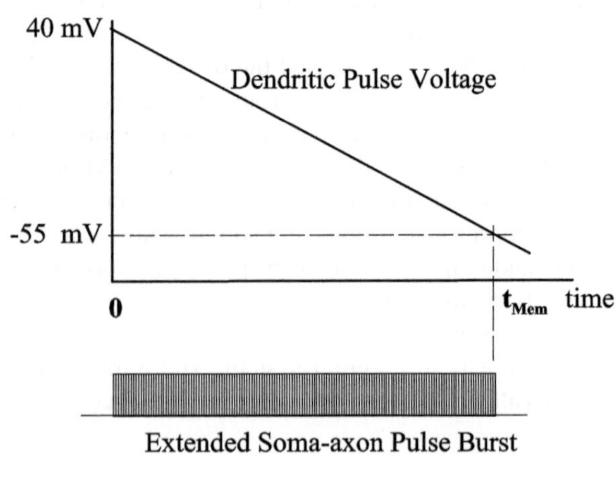

Fig. 3.5 Extended trigger produces an extended pulse burst

To emphasize, Fig. 3.5 shows a dendritic pulse for $\beta = 0.05$ (If $\beta = 0.05$ then Time(ms) = $t_{Mem} = 1.9$ s). As the voltage lowers into a threshold region below −55 mV, the STM neuron will lose its ability to self-trigger and thus its memory.

Electrically, a STM neuron is a pulsating version of a "*one-shot*" in digital design, presenting true output for a given time and then going false.

Synapses

The brain, it is estimated, contains far more synapses then neurons, perhaps 500 trillion (1 trillion $= 10^{12}$); This averages to about 10^3 synapses per neuron, some more, some less. Synapses are numerous, so it is reasonable to expect that they are not all used at once and that they are not all exactly alike.

A *synapse* is a purposeful gap between a bouton at the terminations of an axon, and a receptor, usually located on a spine of a dendrite. This gap is called the *synaptic cleft*; it is typically about 20 nm wide (20×10^{-9} m). Neurons generally do not touch each other physically. Figure 3.6 is an artist's concept of a synapse. Note that the term *presynaptic* refers to the side that contains the bouton; *postsynaptic* refers to the side that contains the receptor.

Pulse bursts can be triggered by *excitatory neurotransmitters* within *excitatory synapses.* Also possible are *inhibitory neurotransmitters* and *inhibitory synapses* that slow down and perhaps stop the propagation of dendritic pulses. Excitatory and inhibitory neurotransmitters may be fast acting or slow acting.

Excitatory synapses trigger the propagation of one or more pulses, and inhibitory neurotransmitters have the power to stop the propagation of pulses. The chief purpose of an excitatory neurotransmitter is to carry a one-way signal from the bouton of a given neuron to the receptor of another neuron. Neurotransmitters, like any molecule in an ionic solution, become positive ions, and once released, they drift and diffuse across the cleft to perform their function.

Intuitive Models for Triggering by Neurotransmitters

Under the simplest of physical models, positive excitatory neurotransmitter ions are ejected from vesicles in boutons because they are repelled by the positive pulses that arrive at the tips of an axon. Neurotransmitters then drift and diffuse away from their presynaptic source to shower a postsynaptic receptor. Neurotransmitter ions may cling briefly to their receptor, partly because the receptors hold a negative rest voltage of about −70 mV, which will attract positive ions.

Once a pulse is triggered, it begins to propagate along the dendrite away from the receptor. The positive pulse also exerts a repulsive force on the positively charged neurotransmitters and will push them away. Most often they soon return to trigger another pulse.

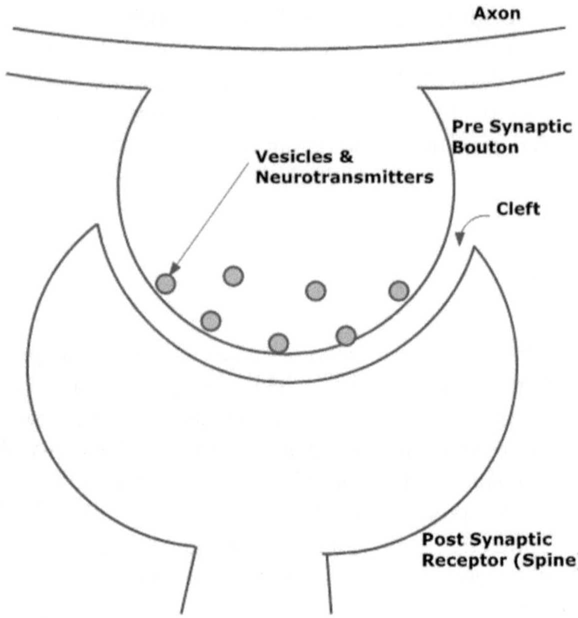

Fig. 3.6 Idealized synapse (not drawn to scale)

Closer to the soma (the body of a neuron), inhibitory neurotransmitter ions may contact and possibly surround a short segment of a branch, with the effect that they deactivate portions of the membrane. Converting the membrane from active to passive would soon attenuate a propagating pulse. An alternative possible mechanism is that inhibitory neurotransmitters may provide internal electrons that tend to discharge neural pulses. For either mechanism, the result is a significant attenuation of propagating dendritic pulses.

There are many varieties of neurotransmitter-receptor combinations, some fast-acting, some slower. A majority of neurotransmitters for neural communications seem to be entities similar to those of *glutamate* (GLU) associated with *excitatory* molecules and *gamma amino butyric acid* (GABA) associated with *inhibitory* molecules.

Excitatory Ions

An intuitive way to visualize an excitory receptor is suggested in Fig. 3.7. Initially, sensitive particles within the membrane are held together by the strong intrinsic electric field within the membrane. But as neurotransmitter ions randomly approach a receiver they disrupt the alignment of membrane molecules. This triggers the membrane and permit sodium ions to begin an action potential, energized in the usual way by sodium, potassium, and other ions.

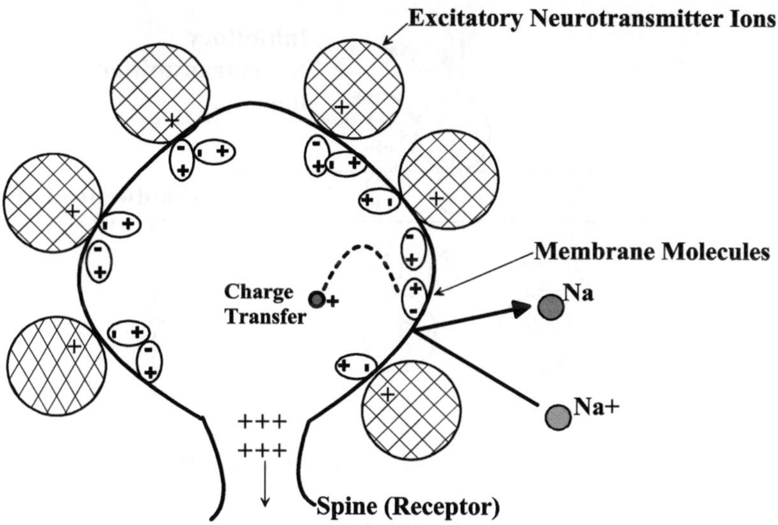

Fig. 3.7 Excitory neurotransmitter ions result in membrane triggering

Eventually a restored negative potential within the presynaptic boutons attracts the free positive neurotransmitter ions and encourages them to "dock" near where they came from. The return of excitatory neurotransmitters has given us the curious word "*reuptake*."

Inhibitory Ions

Figure 3.8 intuitively visualizes larger neurotransmitter ions that uniformly cover a significant area of surface, including receptors of a dendritic segment. The insulating influence of inhibitory neurotransmitter ions is sufficient to prevent the membrane from regenerating a pulse, so pulses attempting to pass through die out rapidly.

When a segment of dendritic membrane is blocked, attenuation is significant because of resistance *R* and conductance *G*, as suggested in the figure. A single pulse that enters a passive segment will soon be too weak to trigger its own regeneration; consequently propagation will stop.

Inhibitory ions must stick long enough to block propagating pulses. Attached inhibitory ions tend to be repelled by dendritic pulses and also by thermal activity; so they, or parts of them likely are attracted back home once their homes return to a negative rest potential.

Circuit Elements for Synapses

The physical model delineated above is that a pulse burst arrives at a bouton and results in a release of neurotransmitters. These in turn trigger pulses in a dendritic receptor.

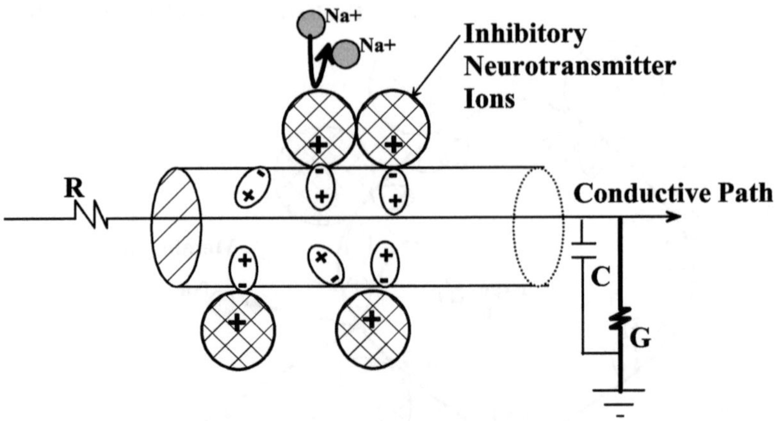

Fig. 3.8 Inhibitory neurotransmitter ions tends to inactivate a membrane

This process can be approximated in a circuit model by viewing a *general synapse* as a one-way *transconductance amplifier* that produces a packet of charge whenever a pulse arrives. Thus, if ten pulses arrive, there will be ten packets of charge in this model; this may generate ten postsynaptic pulses, although there may be situations involving higher membrane capacitance, as below, in which fewer dendritic pulses are generated. This is because some of the packets of charge are employed in raising membrane voltage to a threshold for triggering.

Signals do not usually flow backwards into a bouton. Partly for this reason a one-way transconductance amplifier is a reasonable way to model a *synaptic contact*, a symbol for which is shown in Fig. 3.9. Transconductance G, high load resistance R_L, and low capacitance C_L are available to optimize the contact. So a general synapse is modeled as a transconductance amplifier that transmits one packet of charge for each voltage pulse applied to its input. Under ideal conditions of ideal low capacitive loading at the receptor, each pulse triggers a postsynaptic pulse and is simply transmitted along the destination dendrite.

Weak Synapses, Single Pulses

As opposed to a general synapse, a *weak synapse* is occasionally required that stimulates a single pulse in its dendritic receptor, here denoted by *Pulse*(*1*). A single dendritic pulse is often necessary for precise timing purposes. Such contacts are postulated to be accomplished physically by excitatory neurotransmitters drifting and diffusing through a *small area* cleft to a postsynaptic receptor. Once repelled by the first dendritic pulse, they escape from the cleft so that only one postsynaptic pulse results.

Figure 3.10 proposes an intuitive structure that depends on geometry to encourage a single pulse. Neurotransmitters are emitted as usual, enough to trigger a

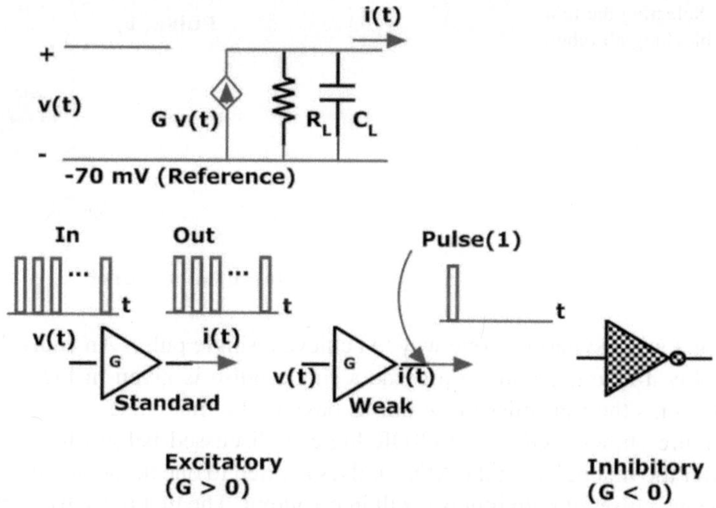

Fig. 3.9 Synaptic contact models; excitatory, excitatory producing a single pulse, inhibitory

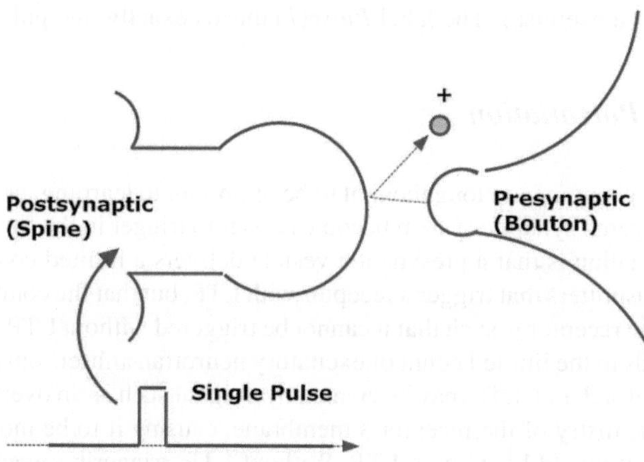

Fig. 3.10 Weak synapse for a single pulse

neural pulse in a receptor. The positive charge in the receptor pulse repels the neurotransmitters out of the region so they cannot return. They do not return because the geometry is such that they are repelled out of the synaptic cleft.

This is in contrast to a synapse that traps neurotransmitters within the cleft so that they oscillate back and forth to give a burst of pulses. This oscillating process terminates when the presynaptic region again returns to a negative voltage, thus attracting positive neurotransmitter ions away from the postsynaptic region.

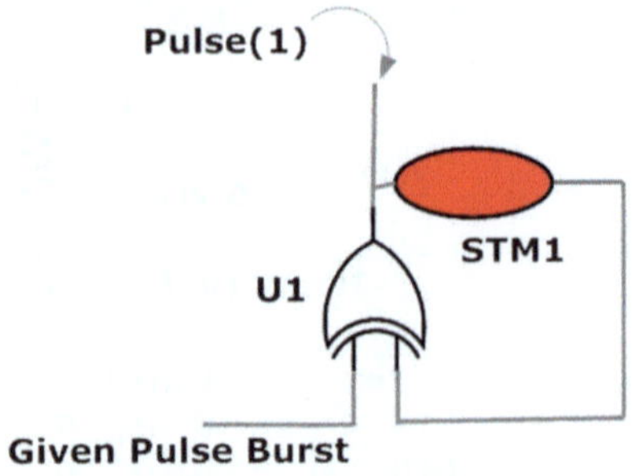

Fig. 3.11 Selecting the first pulse and blocking all others

Having a weak synapse is one way to achieve a single pulse. An equivalent way is to employ a neural circuit to provide a single pulse is given in Fig. 3.11. This circuit prevents the transmission of pulses beyond the first pulse.

The figure shows XOR logic (XOR logic is discussed below) to prevent any pulses after the first pulse. In the XOR, pulses applied to one input pass through, but simultaneous pulses at both inputs result in no output. The first pulse triggers a STM neuron STM1 whose several output pulses are timed to balance each additional pulse in the burst to prevent the transmission of pulses beyond the first. Other less tricky methods of producing a single pulse are possible using enabled AND and NOT gates, described below, which makes pulse timing less critical (the design of a circuit is left as a exercise). The label *Pulse(1)* means exactly one pulse.

Long-Term Potentiation

This refers to a phenomenon long thought to be important to learning, under which a previously triggered synaptic system becomes easier to trigger in the future [3]. One intuitive explanation is that a presynaptic vesicle delivers a limited count of excitatory neurotransmitters that trigger a receptor with LTP, but that the configuration of the postsynaptic receptor is such that it cannot be triggered without LTP. That is, the neuron responds to the limited count of excitatory neurotransmitters only if LTP has been created beforehand. LTP may be created by a signal such as an overvoltage that changes the chemistry of the receptor's membrane, causing it to be more sensitive than otherwise it would be without LTP. Without LTP, triggering would require a full bundle of excitatory neurotransmitters.

LTP Circuit Model

LTP can be modeled in circuit terms as *trapped static charge* that has been permanently placed into the membrane capacitance of a receptor. The receptor with LTP must be able to be queried, so the connecting synapse is modeled as a regular synapse as above, that is, a transconductance amplifier that transmits one packet of charge for each voltage pulse applied to its input.

Trapped static charge across the membrane capacitance of a postsynaptic receptor is equivalent to an *initial voltage* across the capacitor as depicted in Fig. 3.12. Receptor geometry and parameters are assumed to be such that membrane capacitance is a little larger, so that without LTP it cannot be triggered by a single pulse. But with LTP a single pulse is able to trigger the receptor. LTP makes a receptor easier to trigger because not as much additional charge is required to build up capacitive voltage to a threshold for triggering.

For example, the neural gate in the figure can be triggered by a single pulse from a weak synapse, instead of the usual burst of several pulses. That is, the capacitance value is such that without LTP, activation would require several pulses, but with LTP it only requires one pulse.

Unlike ordinary capacitive charge, this charge is trapped and does not readily leak away via membrane conductance. Hidden in the model are mechanisms that maintain a trapped charge. LTP is proposed to achieve elements of long-term memory, needed in subsequent chapters, although it is not the only proposal for such elements.

Circuit Elements for Back Propagation and Charge Storage

When a smaller diameter dendrite physically connects to a larger diameter dendrite or soma, there is added membrane capacitance as suggested by the symbol in Fig. 3.13. The higher membrane capacitance C can integrate the charges being delivered by impinging pulses and thus can accumulate voltage. When incoming pulses have charged the capacitance sufficiently, its voltage reaches a threshold that triggers an action potential in the larger body.

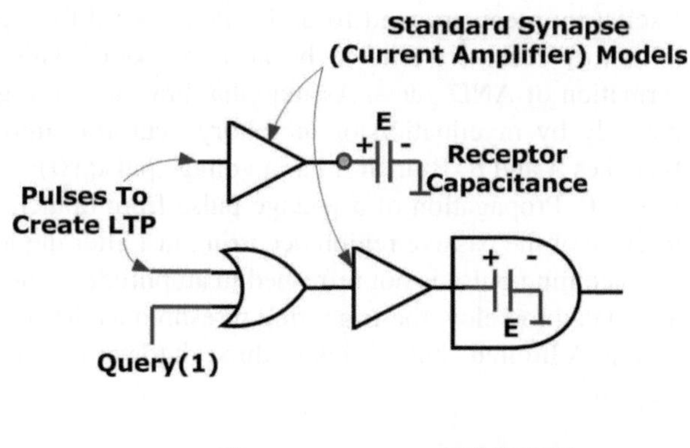

Fig. 3.12 Synaptic contact with long-term potentiation

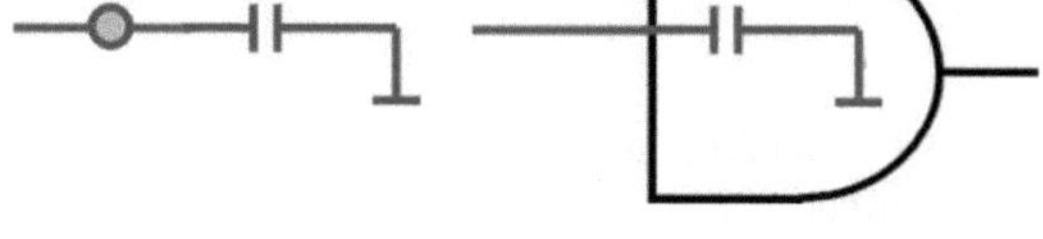

Fig. 3.13 Physical contact to a larger body

Simulations suggest that a capacitive termination, such as at a soma, results in back propagations away from the capacitor. Based on simulations, back propagations are fairly common, although their existence depends on the amount of capacitance involved. Back propagation occurs because the capacitance holds charge sufficiently long so as to retrigger the connecting dendrite at a time when it has recovered to be triggered again, sending a propagation back away from the soma.

At the input tips of a dendrite, there are no reflections, apparently because there is no extra capacitance. According to simulations pulses simply disappear. So back propagations do not return again. A similar situation of pulse elimination occurs at the termination tips of an axon.

Dendritic Logic

The neural gates described next occur in dendrites and are illustrated with dendritic junctions. Although difficult to visualize, it might be possible for such logic to be generated without an obvious junction, assuming the equivalent of a junction occurs because of inhomogeneity in a dendritic segment. Also it might be possible for close-spaced synapses to produce signals that combine logically in their dendrites, depending on geometry. However, this is considered dendritic logic since it occurs in dendrites not in synapses. Several forms of neural logic are possible, but in this section the focus is on dendritic logic.

Dendritic AND Gate

Excitatory receivers tend to be located toward the tips of the dendrites, whereas inhibitory receivers tend to be near the soma. This, it turns out, facilitates the formation of AND gates. Assume that branch Y in Fig. 3.14 is pacified somehow, possibly by myelination or inhibitory neurotransmitters near the intersection of branches A and B. Branch A has a voltage pulse $v(t) > 0$, while branch B has none, $v(t) = 0$. Propagation of a voltage pulse from branch A to the output Y is stopped because of the passive region occurring just after the junction. This occurs because the incoming pulse is not refreshed in amplitude in the passive region, so voltage is attenuated to below the triggering threshold as the pulse emerges from the passive region. A limited charge trickles through toward Y, not enough to trigger the active portion.

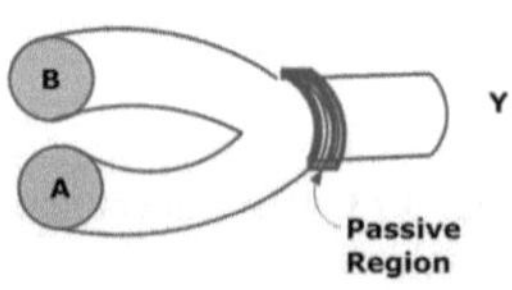

Fig. 3.14 Dendritic AND gate concept

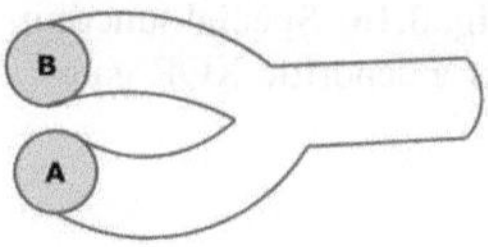

Fig. 3.15 Simple junction for the dendritic OR gate

But if both branches A and B have voltage pulse inputs, A and B work together, and there is additional current (I) to charge the passive region to a greater voltage peak (because of the basic formula $\Delta V = I\Delta t/C$). There can be enough voltage to trigger the right most active region, assuming pulses from A and B impinge at the same time.

Define A, B, and Y to be Boolean variables which are false for no action potentials and true if there are neural pulses. If A is true and B is false, Y is false, as described above. Similarly, if B is true but A is false, Y is still false, because B cannot by itself send enough charge through a passive segment to trigger segment Y. But if both A and B are true, then Y is true, because by working together, sufficient charge accumulates in the inhibited segment to continue the propagation of the dendritic pulse. The result is an AND gate:

$$Y = AB. \tag{3.4}$$

The length of the passive region has to fall within a certain range; if it is too short, its capacitance will not stop pulses from a given branch; if too long, it will stop even two combined pulses.

It is emphasized that in order for the above AND gate to work, the pulses must arrive at about the same time.

Dendritic OR Gate

An OR gate assumes no inhibitors in either branch. In this case, the soma will be activated for input to A or input to B, or if both inputs are active as in Fig. 3.15. This is an uncomplicated physical junction of dendritic branches. In this case, the Boolean logic is

$$Y = A + B. \tag{3.5}$$

The OR gate is not dependent upon pulse coordination since pulses arriving at differing times are naturally passed through.

Exclusive OR and NOT Gates

Dendritic XOR Gate

Under certain fairly easy-to-achieve conditions, the OR will behave like an exclusive OR, denoted as XOR, meaning that any one signal will be transmitted, but that two applied together will not be transmitted. This has been shown by simulations [4].

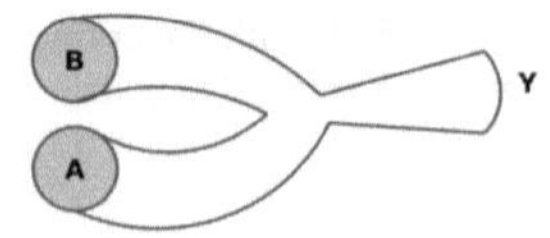

Fig. 3.16 Special junction for a dendritic XOR gate

In the above model, this behavior depends of the fact that two colliding neural pulses generally annihilate each other, in which case nothing is transmitted. This is found to occur if the average capacitance at a junction is reduced slightly.

Physically a slight reduction of capacitance may be accomplished by partial myelination or by a judicious sprinkling of inhibitory neurotransmitters, although this would also reduce charging currents. Charging currents could be compensated by the local structure of the membrane, that is, it could be slightly thinner to increase charge penetration or the local ionic concentrations surrounding it could be locally increased very slightly. Capacitance can also be reduced slightly by a local reduction in diameter as pictured in Fig. 3.16.

The result is that one pulse will penetrate, but two pulses will not. When two pulses arrive simultaneously, they collide; colliding pulses annihilate, as clearly demonstrated by simulations. Thus

$$Y = A \oplus B. \tag{3.6}$$

It must be emphasized that in order for the above XOR gate to prevent the passage of two pulses, the pulses must arrive at about the same time. The need for pulse timing is reduced in enabled logic explained later.

XOR NOT Gate

The XOR function is such that it can be modified to be a NOT gate. This is because of its theoretical equation:

$$Y = AB' + A'B. \tag{3.7}$$

If B is made true (=1), then B' or *NOT* (B) is false (=0), and so by algebraic reduction:

$$Y = A'. \tag{3.8}$$

Using neural pulses, the XOR will function as an inverter only under certain conditions. When pulses are applied to one input, concurrent auxiliary pulses may applied simultaneously to the other input to make the output zero. But if at any time only auxiliary pulses are applied to one input and not the other, the output will be true, with pulses emitted. Pulse coordination is less important using *enabled* logic presented later.

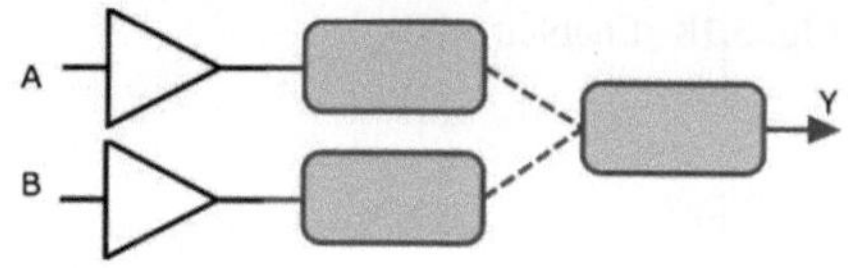

Fig. 3.17 Two synapses can constitute an OR gate

The Possibility of Synaptic Logic

Synapses make contact to spines on a given dendritic branch. The possibility of synaptic weights has been promoted [5, 6] but is discounted in this book, owing to a membrane that either pulses or it does not. And when it pulses, it provides a standard pulse. However, standardized pulses from adjacent synapses (on spines) may very well proceed to undergo dendritic logic.

Synaptic logic, which the author considers to be dendritic logic, might be said to be occurring when adjacent active spines provide pulses to dendritic membrane that is either active or inactive. If membrane is active, an OR gate results with the circuit model in Fig. 3.17. If the membrane is passive for a short distance, a dendritic AND gate might result. Note that these are in fact examples of dendritic logic.

To an electrical engineer, the figure is reminiscent of hard-wired diode logic, but the underlying physics is completely different. The signals are bursts of pulses. Either input *A*, input *B*, or both can initiate a burst of pulses in the dendrite to produce a propagation at *Y*. Note that synapses send pulses one way and thus avoid back propagations.

Enabled Logic in Dendrites and Soma

Enabled logic depends on applying charge to an input capacitance until a triggering threshold is reached. In particular, an AND gate is achieved if one input supplies half of the required charge, and the other input supplies the other half to trigger a pulse burst. Enabled logic does not required that pulses arrive simultaneously, so in this sense it is simpler. However, pulses must still arrive within a certain time frame or existing charge will dissipate. A place to accumulate charge is required, often taken to be the capacitance of the larger soma in a neuron; there is only one soma per neuron, so there will be fewer enabled gates per neuron.

Enabled logic is akin to artificial neural network concepts under which (usually DC) inputs are added together linearly to reach a threshold. However, the weightings of the inputs are usually fixed in enabled logic, since pulse bursts from active membrane generally have fixed waveforms. Specialized neurons may produce differing counts of pulses in a burst, and even differing waveforms, however, so more exotic forms of weightings may be possible.

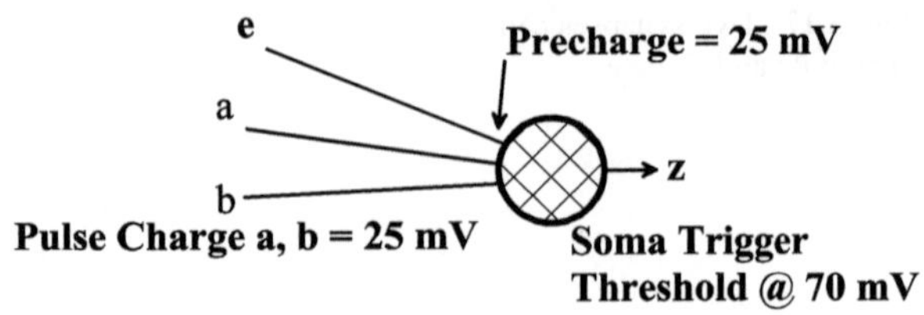

Fig. 3.18 Enabled AND

Enabled AND

Figure 3.18 illustrates the AND using assumed values for the purposes of convenience (From the neural pulse given earlier, a realistic threshold is roughly 15 mV above the rest voltage 70 mV).

This is a three-input AND gate with "*e*" used as an optional enable that contributes 25 mV. The soma is triggered only if inputs "*a*" and "*b*" provide sufficient charge to reach the threshold assumed to be at 70 mV in this illustration. The count of pulses must be able to build up about 25 mV from *a* and about 25 mV from "*b*". Then $z = e(ab)$. Enabled logic does not depend as much on pulse coordination.

Enabled OR

The OR results in the above figure by requiring a contribution of at least 50 mV for each input (a, b). Then $z = e(a + b)$

Enabled NOT Gate

Without the availability of negative voltage as commonly used in artificial neurons, one wonders if a NOT gate is indeed possible biologically. To investigate NOT gates, consider the more complicated situation of enabled logic depicted in Fig. 3.19. A precharge in the soma could be 50 mV positive, for example, and it might be defined to trigger at 70 mV. Thus another 50 mV from incoming pulses would enable dendritic logic.

The code in this diagram is as in Table 3.1.

Assume branch "*b*" is receiving dendritic solitons prior to the NOT operation and that precharging to 50 mV has been completed via branch "*e*." Assume branch "*a*" in the left-most figure has impinging pulses. The propagation of solitons to the soma is stopped at branch *y*, because, linked to neural input "*a*" are inhibitory neurotransmitters for branch "*y*." This linking is denoted by the primes in the above figure, that is, "(+)' and (−)'." The action potential applied as in input to branch "*b*" is also stopped for similar reasons. Effectively there is an AND gate at *y*.

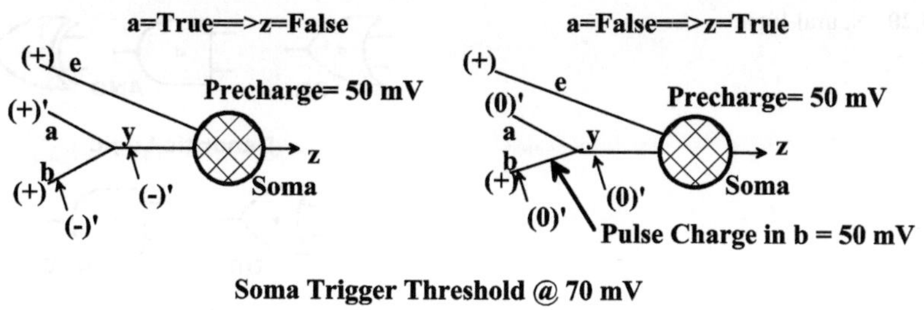

Fig. 3.19 NOT gate concept for a neuron

Table 3.1 Neurotransmitter codes

(+)	Presence of excitatory neurotransmitters
(−)	Presence of inhibitory neurotransmitters
(0)	Absence of neurotransmitters

But there is no chance of signals "*a*" and "*b*" producing an output, since signal "*b*" does not pass through to reach junction *y*. Thus the output of axon "*z*" is false (no output pulses) when the inputs going to branch "*a*" are true.

If branch "*a*" is not excited, as in the right-most figure, the inhibitory neurotransmitters from the primed input are likewise not applied. Branch "*b*" is now free to propagate an action potential. This signal is not stopped by branch "*a*" where an action potential merely propagates harmlessly back down branch "*a*" and, most importantly, also toward the soma. Thus, when the inputs to branch "*a*" are false, the soma output is true. What is accomplished is a NOT gate using a neuron, assuming a linked combination of pulses to "*a*" and inhibitory neurotransmitters to "*b*" and "*y*."

The external neuron driving branch "*a*" must also provide inhibitory inputs to branches "*b*" and "*y*." The inputs to branches "*e*" as well as "*b*" may be obtained from an external neuron. Branch "*e*" serves as an enable, as does the input to branch "*b*." To summarize using Boolean notation:

$$z = e(a')b,$$

z, e, a, b are Boolean signals and a' means NOT(a). To enable it, let $e = b = 1$ so that $z = a'$. Note that no dendritic gates are involved. It also may be noted that the timing of the precharge due to branch "*e*" is not critical; but it must be present to have a valid output. The input to branch "*b*" is similarly not critical. These auxiliary signals may be stopped only after operations are completed. For enabled neural logic, signals need be present only during a certain relevant time frame.

Focusing on basic neural components as above is convenient when explaining the logical properties of a neuron. Nevertheless there are several proposed ways to achieve a complete set of Boolean logic [7–12].

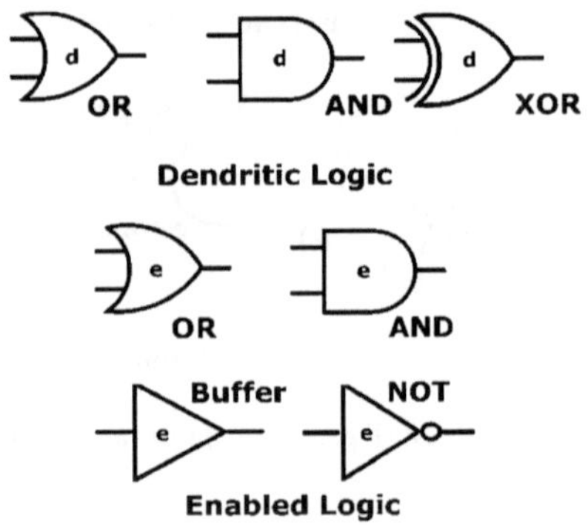

Fig. 3.20 Neural logic symbols

Generalized Neural Logic

Figure 3.20 summarizes symbols for neural logic. Generally the inputs and outputs are assumed to be neural pulses. The little letters within the symbols are not always employed if the context of the logic circuit is clear; **d** refers to dendritic; **e** refers to enabled. Delay is not considered a logic element, although it is part and parcel of dendritic systems and is important to a logic system. Dendritic AND, XOR require that pulses arrive simultaneously.

Dendritic logic is subject to back propagations such that signals flow may flow back from output to input. OR, XOR permit an easy flow backwards. AND will prevent back propagation assuming two-input, one-output geometry and uniform diameters of dendrites.

Enabled logic may have an axon that conducts the output pulse; axons resist back propagation. There is no enabled XOR using a single neuron, although one might be possible by considering two layers of neurons with special conditions on numbers of pulses. It is not clear that suitable conditions are possible without excessive complication. The buffer and NOT symbols are similar to the synaptic symbols; the context of usage must be clear or the designator **d** and **e** must be used.

Conclusions

Two fundamental forms of neural logic are presented: dendritic logic and enabled logic. The concept of weighted synapses, used for the engineering of artificial neural networks is not used here, but rather "synaptic" logic is recognized to be the same as dendritic logic. Dendritic logic depends on the timing of pulses propagating along dendritic branches, since pulses must arrive nearly simultaneously to accomplish the desired logic, which is limited to combinations of AND, OR, and XOR gates.

In contrast, enabled logic depends on an ability of a membrane to accumulate voltage to reach a threshold for triggering. Voltage is accumulated from pulses that may add as required to produce AND, OR gates. Enabled NOT gates are also possible when inhibitory neurotransmitters are considered. Enabled logic is not as dependent on timing. Instances of dendritic logic may occur thousands of times in dendrites, whereas enabled logic is generally limited to active regions that can store charge, usually a soma.

Several new symbols for circuit elements have been introduced. The usual symbols for capacitance and resistance are used for membrane charge, membrane conductance, and channel resistance. Controlled current source symbols are used for membrane charging, while control circuits as in the appendix freely employ independent voltage source symbols and ideal diodes.

A capacitor symbol represents a step in membrane capacitance, which happens when a smaller dendrite physically connects to a larger body, such as a soma. This is important because capacitance in the larger body accumulates charge, and also, according to simulations, may trigger a back propagation of pulses.

Pulses, akin to solitons with fixed waveforms are produced by a continuously active membrane, and typically flow back and forth within dendrites, but not back into presynaptic regions.

A synapse transmits signals one way, from the presynaptic (bouton at an axon termination) region to postsynaptic (receptor on spine) region. A excitatory synaptic contact is denoted by a transconductance amplifier of the non-inverting type such that signals flow from input to output. This amplifier injects positive charge to excite neural pulses. An inhibitory synaptic contact is denoted by an inverting transconductance amplifier. This amplifier injects negative charge to dampen neural pulses. These amplifiers are denoted by triangular amplifier symbols that point in the direction of signal flow.

A postsynaptic receptor can accumulate LTP, denoted by a capacitor with initial static charge, indicating a receptor region that is precharged so that it is easier to trigger. LTP is thought to be important to long-term memory and learning.

An STM neuron is denoted with an oval symbol. STM neurons are modeled as having a shortfall of internal ions in its dendrites and also by an insulated membrane so that a longer dendritic pulse results; this enables the triggering of an extended burst of pulses at a maximal rate to be sent down the axon. Burst durations vary and depend on internal parameters.

Symbols for neurotransmitters were introduced in conjunction with enabled NOT gates: (+) for excitatory, (−) for inhibitory, and (0) for no neurotransmitters. Neural-synaptic circuit elements are the subject of simulation experiments described in the appendix, and will be used to model long-term memory, simulated qubits, and various systems such as cue editors and recall referees as in subsequent chapters.

References

1. Burger JR (2009) Human memory modeled with standard analog and digital circuits: inspiration for man-made computers. Wiley, Hoboken, NJ
2. Kandel ER (2006) In search of memory: the emergence of a new science of mind. W. W Norton & Co, New York

3. Bliss T, Collingridge GL, Morris RGM (2004) Long-term potentiation: enhancing neuroscience for 30 years. Oxford University Press, Oxford
4. Burger JR (2010) XOR at a single point where dendrites merge. arXiv:1004.2280 [cs.NE] April 13
5. Haykin S (1994) Neural networks: a comprehensive foundation. Macmillan, New York
6. Arbib MA (2003) The handbook of brain theory and neural networks. MIT Press, Cambridge, MA
7. Stuart G et al (2008) Dendrites, 2nd edn. Oxford University Press, New York, p 328, 351–399, 225–250, 421–440
8. Poirazi P, Brannon T, Mel BW (2003) Pyramidal neuron as two-layer neural network. Neuron 37:989–999
9. Koch C, Segev I (1989) Methods in neuronal modeling: from ions to networks. MIT Press, Cambridge, MA
10. Segev I, Rinzel J, Shepherd G (1995) The theoretical foundation of dendritic function: selected papers of Wilfrid Rall with commentaries. MIT Press, Cambridge, MA, Ch. 1, 2, 8
11. Mel BW (1994) Information processing in dendritic trees. Neural Comput 6:1031–1085
12. Fromherz P, Gaede V (1993) Exclusive-OR function of a single arborized neuron. Biol Cybern 69:337–344

Self-Study Exercises

(If you desire to keep a record of your efforts, please show a couple of logical steps leading to the answers.)

Neural Pulses

1. Apply the simplified neural pulse model to plot a neural pulse assuming:
   (a) $\alpha = 110, \beta = 86.66$; Determine pulse width at −55 mV
   (b) $\alpha = 110, \beta = 1$; Determine pulse width at −55 mV
   (c) $\alpha = 110, \beta = 0.01$; Determine pulse width at −55 mV
   (d) Derive a formula for time for a pulse to decrease from +40 to −55 mV in terms of β. (ANS. $t_\Delta = 1 + 95/\beta$)
2. Using 15 ms/cm
   (a) How long is a line for a delay of 90 ms? (ANS. 6 cm)
   (b) How long is a line for a delay of 600 μs? (ANS. 400 μm)
3. Assume a burst of neural pulses caused by an excitatory neurotransmitter to a certain synapse.
   (a) Describe briefly what causes the burst to end. (ANS. Presynaptic region goes negative, tending to attract back home the positive neurotransmitter ions; they are also encouraged to leave by back propagating positive pulses.)

(b) Devise two ways to pass only the first pulse of a burst using an AND gate (HINT: Methods given in this book includes the XOR and another, the AND along with STM neurons)

Short-Term Memory Neurons

4. An STM neuron has evolved to provide a pulse burst of 1,500 ms. Assume that a standard neural pulse can occur every 5 ms.
   (a) What is the frequency in a standard pulse burst? (ANS. 200 Hz)
   (b) How many pulses are emitted by this STM neuron? (ANS. 300)

Logic Involving Synapses

5. Illustrate logic for $A + (B + C) + D$ (HINT: transconductance amplifiers are involved).
6. Consider two neural signals E and F. Produce a sketch of a timing diagram to illustrate the following, assuming that E arrives at time 0, and F arrives with a delay of 1 ms
   (a) Assume an OR gate: $E + F$ (HINT: Show a subsketch of E and F separately)
   (b) Assume an AND gate: EF

Long-Term Potentiation

7. LTP of say, 10 mV can make a neural component easier to trigger. Assume triggering occurs at 15 mV:
   (a) What is the voltage accumulation to trigger? (ANS. +5 mV)
   (b) What would the size of the pulse have to be with 5 mV of LTP? (ANS. +10 mV)

Capacitive Loads and Back Propagations

8. A capacitive load such as that at a neural soma produces back propagation in dendrites. Assume that when a forward moving pulse meets a back moving pulse, that a collision occurs that annihilates both pulses.
   (a) If pulses arrive from a synapse at a rate of 10 Hz, and travel down an active branch to a soma, what is the rate of pulses arriving at the soma? (ANS. 5 Hz)

(b) If a back propagation arrives at a synaptic receptor or spine that is no longer active in producing pulses, what happens to the back traveling pulse. (ANS. It goes to the end of the dendrites and vanishes with no further reflections, because there is no capacitive load there.)

9. State the effects of a back propagating pulse on the following:

   (a) Excitatory neurotransmitters at a receptor (ANS. repels positive ions away from the excitatory receptor)
   (b) Inhibitory neurotransmitters around a dendrite (ANS. helps repels positive ions away from the dendrite)
   (c) Can back propagation ever stop a forward moving pulse? (ANS. Yes, colliding pulses annihilate)

Boolean Logic

10. Provide a logic schematic to give the following Boolean function: $Z_1 = AB + C + B'$ (HINT: Consider reducing the function to a simpler form $Z_1 = A + B' + C$)
11. Provide a logic schematic to give the following Boolean function: $Z_2 = (A + B)B' + CD$ (HINT: Consider reducing the function to a simpler form $Z_2 = AB' + CD$)

Dendritic Logic

11. Illustrate the following with sketches of dendrites. (A) If pulses are applied to the output, predict what might appear at the inputs, (B) If pulses are applied to one input, predict what might appear at the other inputs.

    (a) OR gate (ANS. Back propagating pulses on inputs; back propagating pulses on other input)
    (b) AND gate (ANS. No Pulses on inputs; back propagating pulses on other input)
    (c) XOR gate (ANS. Back propagating pulses on inputs; back propagating pulses on other input)

12. Comment of the timing of pulses for the following dendritic logic gates:

    (a) OR gate (ANS. Pulses may come at random)
    (b) AND gate (ANS. Pulses to the inputs must come simultaneously to pass a signal)
    (c) NOT gate using XOR (ANS. Pulses to the inputs must come simultaneously to make output be at rest)

Enabled Logic

13. Explain how enabled neural logic works as in this chapter?
    (a) Explain the two-input enabled OR gate (HINT: 50 mV is provided for enabling, and 50 mV or more from any input will trigger the subject soma, assumed to trigger at 70 mV)
    (b) Explain the two-input enabled AND gate (HINT: 50 mV is provided for enabling, and 25 mV from each input will trigger the subject soma, assumed to trigger at 70 mV)
14. Provide a table that list all:
    (a) Dendritic logic elements
    (b) Enabled logic elements
    (c) Other types of circuit elements (HINT: Refer to symbols mentioned in this chapter)
15. Write out a truth table for:
    (a) The AND function, that is, AND(a, b)
    (b) The OR function, that is, OR(a, b)
    (c) An exclusive OR, that is, XOR(a, b)
    (d) An exclusive NOR, that is, XNOR(a, b)
    ANS.

a	b	AND(a, b)	OR(a, b)	XOR(a, b)	XNOR(a, b)
0	0	0	0	0	1
0	1	0	1	1	0
1	0	0	1	1	0
1	1	1	1	0	1

    (e) Locate a NOT function in the table (ANS. Let $a = 1$ and mark the XOR versus b)
16. Provide a diagram of the simplest possible NOT gate using a neuron and neurotransmitters, using the symbols of this chapter. (HINT: Refer to Fig. 3.18.)
17. What extra requirements are there for the neural NOT gate?
    (a) Using dendritic XOR gate (ANS. an auxiliary burst of synchronized pulses is necessary during operation)
    (b) Using enabled gate (ANS. inhibitory neurotransmitters and an auxiliary burst of pulses is necessary during operation)

Chapter 4
Long-Term Memory, Simulated Qubits, Physical Qubits

Introduction

The five senses transmit a large number of signals to the brain, and after they are processed, they present an image in conscious short-term memory that is composed of fundamental attributes, that is, specific colors, shades, shapes, edges, tones, and so on. The existence of an particular attribute in a given image may be indicated by a simple true (present) or false (absent). Each image in long-term memory is thus a binary record. Additionally, each attribute of each image in long-term memory will have to have links to conscious short-term memory, so that recall is possible.

Memory, be it short and long term, depends on bistable memory elements, one for each attribute, each holding true or false. This is in contrast to other fanciful concepts. For instance the grandmother neuron is supposed to hold an image of one's grandmother, we are told. But it is difficult to imagine how a grandmother neuron would be searched among the maze of other such neurons, and then transferred into conscious short-term memory. To avoid such practical problems, this book stays with the concept of one neuron per attribute in what has been termed *distributed memory* [1].

More is known today than say, a decade ago, but much remains unknown about the physical details of human memory. Reasons are, memory neurons have minimal impact on their surroundings. Neurons are efficient with energy compared to solid-state devices, and so taken individually, they do not release very much heat and radiation, making them difficult to trace. They are microscopic; they are tightly integrated with billions of other neurons, and so are difficult to isolate in action. In vivo experimentation on a connected neuron is especially challenging since they are easily rendered inoperative by probing.

Considering possibilities for long-term memory elements, there are differing purposes. A basic element stores a signal for the presence of a given attribute, and returns its information with 100 % fidelity, a simple true or false. Recursive neurons, on the other hand, store a basic logic value, but can return their logic value

J.R. Burger, *Brain Theory From A Circuits And Systems Perspective*, Springer Series in Cognitive and Neural Systems 6, DOI 10.1007/978-1-4614-6412-9_4,
© Springer Science+Business Media New York 2013

with a specified probability, as explained below. This is useful when probabilistic processing is needed, as for certain types of memory search.

An important configuration of a recursive neuron is the controlled toggle, examples of which are delineated below. A controlled toggle is also a basic memory element, but must be toggled from its current value to its new logic value, and cannot simply be set, or over written as though it were a random access memory device.

This chapter begins with a discussion of basic memory elements, including long-term potentiation (LTP) as the basis for an element of memory that holds a logic value for an attribute with 100 % fidelity. A recursive neuron may also serve as a basic memory element to hold a logic value for an attribute with 100 % fidelity. So two possible styles of long-term attribute memory are presented below. Also discussed below is the real possibility of a combination of these two types of elements.

Neurons Configured into Elements of Memory

Associative search through the words of long-term memory and related subliminal processing all depend on the availability of practical memory elements. In practice these must store a logic value (a true) for attributes that are needed, and they must have ability to return their information (to be read) by the system in a convenient way. Pursuing these ground rules, this chapter introduces basic memory elements. Two basic mechanisms for long-term memory are proposed: *LTP* and *Recursive Memory*.

Memory Based on Long-Term Potentiation

This phenomenon is thought to be related to learning as well as to long-term memory [2]. One may envision a memory subcircuit model as in Fig. 4.1.

In this model a regular synapse S2 associated with (memory) neuron U2 is pulsed (possibly in a certain way) many times via OR neuron U1. This makes U2 more receptive to being triggered. It is implied that the receptor in question is susceptible to LTP. Subsequently a read signal (R) from a weak synapse results in a weak query consisting of ideally, only one pulse, designated by *Query*(*1*), that, although weak, will still produce a dendritic pulse burst. Without LTP there would be no trigger, since ordinarily the capacitance of the receptor on U2 is such that it needs to be charged by more than one pulse to result in triggering.

Note that this process is modeled with charge from pulses that accumulate in membrane capacitance, as explained in a previous chapter. The geometry and the membrane parameters of the receptor on its spine are such as to result in a slightly higher capacitance that cannot be triggered by a single pulse. To ensure that the LTP

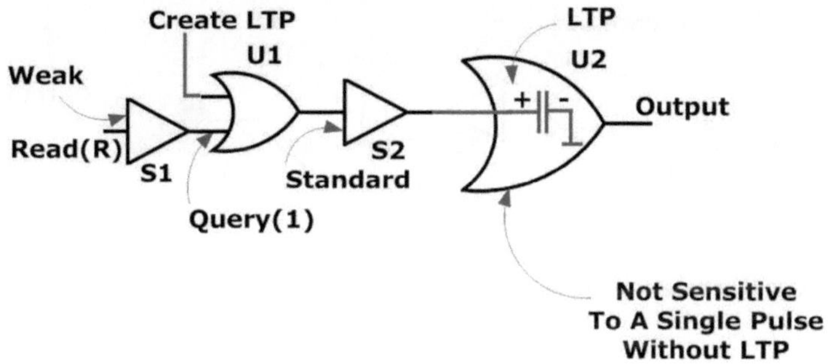

Fig. 4.1 LTP memory circuit model

is accessible to be queried, regular synapse S2 is modeled as a transconductance amplifier that transmits one packet of charge when one voltage pulse is applied to its input. The neuron responds to a single pulse such as forwarded by *Query*(*1*) only if LTP has been created beforehand.

Memorization in this model means creating LTP, which means creating trapped static charge (and storing energy). This concept is vaguely similar to that of dynamic random access memory in home computers, in which a bit of data is stored in the form of charge in a small capacitor. The difference is, dynamic memory requires periodic refreshing every few milliseconds, whereas LTP is assumed to last a very long time without refreshing.

Open questions about LTP are as follows.

1. What makes a receptor susceptible to LTP? It seems that these particular synapses are perhaps special, since it is doubtful that a brain would work in practice if each and every synapse eventually acquired LTP. An intractable situation would be that many attempted queries to an element that is supposed to hold a false, would eventually convert it into a true, if LTP is merely a matter of applying many pulses. To create LTP, there has to be a special signal or code that creates it, perhaps high amplitude pulses as used in the laboratory when studying LTP.
2. Is the time to instill LTP too long? It seems like it would take excessive time to create LTP using the kinds of neural signals that are readily available in a live brain. This is said in view of the observation that major human memories can be formed instantly. For example, emotional experiences often result in instant memorization of times, places, and events, such as destructive enemy attacks, the murder of well-known personalities, and so on. Beyond such unusual circumstances, a lucky few have the gift of photographic memory, implying instant verbatim memorization of massive amounts (books full) of information. Fact is, long-term memorization often occurs instantly.
3. Can LTP actually last for decades without refreshing? Trapped charge in a biological (or any other) environment tends to dissipate over time, partly due to thermal activity, partly due to external radiation such as X-ray or cosmic,

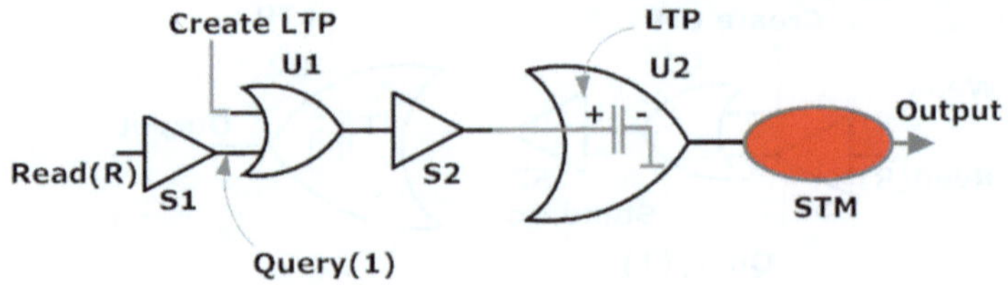

Fig. 4.2 Extending an output signal

depending on how tightly it is held, suggesting that LTP might in fact require refreshing. Certainly it is known that firmly trapped charge in solid-state systems cannot be claimed to last more than about 5 years.

Stretched Memory Signals Using a Burst Stretcher

Short-term memory neurons are plentiful and so may be applied in places other than explicit short-term memory. For example, they may take a neural burst and make it longer, to aid with timing (since all signals ideally must come together at about the same time in consciousness). Once triggered, a short-term memory neuron emits an burst of pulses whose length depends on the chemistry of its dendrite. Figure 4.2 shows how a short-term neuron can stretch the output of a memory element based on LTP, thus making the output signal easier to register in consciousness. This is termed burst stretching.

A stretched signal from a memory element is reminiscent (in the author's mind) of the output of a recursive neural circuit.

Recursive Neurons with a Circulating Pulse

Neurons could theoretically be fed back on themselves to support a circulating pulse, forming a device that is characterized by an efficient electrical pulse that flows indefinitely round a closed circuit until stopped by an outside signal. Since the pulses operate between two given voltage levels, such circuits are multivibrators, which are a type of oscillator that delivers a rectangular-shaped waveform. The model in Fig. 4.3 is typical of a neural multivibrator. U1 is assumed sensitive to a single pulse without a need for LTP. Depending on ancillary logic, a multivibrating circuit may pulsate for a few milliseconds or they may pulsate indefinitely (until power is removed).

Loop delay, which is a variable, is assumed greater than pulse width. Upon receiving a set (S) signal, a single pulse is initiated by a weak synapse denoted by S1. This pulse cycles back to the input via regular synapse S2, and cycles

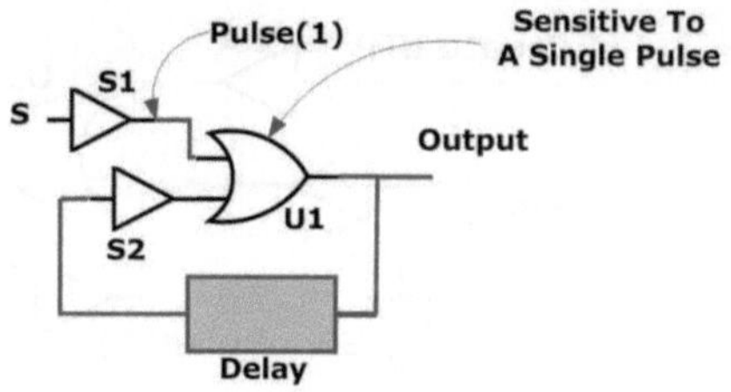

Fig. 4.3 Recursive neural circuit

indefinitely keeping its shape because of membrane activity, and because delay permits the membrane to recover before the pulse appears again. S2 ensures that the pulse propagates clockwise. Once created, a circulating pulse gives a true signal at the output; with no circulating pulse the output defaults to a false signal, that is, rest at about −70 mV.

Delay is affected mainly by the length of the path of the circuit and also by such parameters as membrane capacitance, makeup of the membrane, and local ionic concentrations. Note it differs fundamentally from a ring oscillator of digital technology, which depends on a cascade of inverters connected in a ring. It also differs in that neural membranes produce very much slower propagations than metallic conductors, roughly 15 ms per centimeter in dendrites.

Initially all neurons are at their rest potential (−70 mV). A pulse introduced for cycling reaches up to about +40 mV with a duration of perhaps 2 ms. What is interesting about a recursive neuron is that memorization would be instant and theoretically could last indefinitely. But there are open questions.

The main question is, would recursive neurons operating nonstop, consume excessive energy? Indeed the brain does dissipate significant calories, 10–20 % of all body calories (for a body at complete rest). But losses in series resistance of neurons are low because of the relatively slow rise and falls of a neural pulse, so a pulse does not consume major amounts of energy in series resistance. In this regard neurons are relatively efficient compared to solid-state logic.

However, membrane conductance has losses in proportion to the average voltage across the membrane.[1] On average, losses in the membrane conductance are balanced by a lethargic process of metabolic energy delivery via oxygenated hemoglobin. Recursive neurons, like any other neuron, would take energy, and of course, a lot of energy is available. The presence of a cycling pulse implies an actual reduction in conductive energy dissipation.

Multivibrators and their modifications have important applications, such as simulated qubits and controlled toggling, discussed later in this chapter.

[1] Membrane loss is known to go down during a burst of pulses, since pulsing actually decreases average membrane voltage from −70 mV up to nearly 0. This leaves behind greater amounts of oxygenated hemoglobin so as to explain in part why neural activity gives a brighter region in functional magnetic resonance imaging (fMRI), as mentioned again in a later chapter.

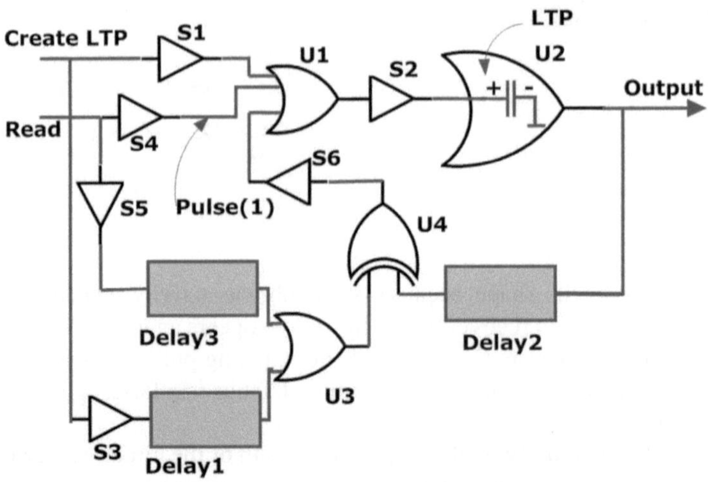

Fig. 4.4 Hybrid long-term memory (S4, S5 weak synapses)

Hybrid Circuit

LTP and recursive circuitry may be combined, and may very well have evolved to provide certain physical advantages, including instant memorization when conditions are right, and memory signals that shut off after a few seconds to conserve energy, but which are easily reactivated. A proposed circuit is illustrated in Fig. 4.4. Note that the physical neural circuitry is fairly simple, but that circuit models often show many technical details including delays.

LTP is created to reside in U2, so that it responds to a single pulse. The signal for its creation flows from S1, U1, and S2. However, the path S3, Delay1, and U3 presents a pulse to dendritic XOR gate U4 whose purpose is to prevent a circulating pulse from being initiated. Delay1 is such that when the two pulses, one recycling from Delay2 and one from U3 meet at about the same time, they give no output from the XOR. So LTP is instilled but with no circulating pulse.

Upon query, if and only if LTP is in U2, multivibration is triggered by a read signal. Note that the query is produced by a weak synapse S4 to have only a single pulse. By the way, the frequency of multivibration is controlled by Delay2.

But the circulating pulse is stopped eventually by a pulse along the path weak synapse S5, Delay3, and U3 to the XOR. Delays3 is such that a prescribed number of cycles through U2 occur but that at some point a pulse emerges from Delay3 and thus from U3. The XOR does not permit an output when two pulses occur simultaneously at the input. The plan is that recursive memory now goes to sleep until triggered by the next query.

This circuit does not take advantage of back propagations. In fact, back propagations are suppressed, and do no harm, by having regular synapses at the inputs of the OR gates.

Other circuits involving an enabled AND gate could be used in place of the XOR and could provide the same sort of function without a need for coordinated timing (devising this circuit is left as an exercise). Hybrid arrangements are not as complex as they look and are quite useful whenever a specific pulse burst is required. Recursive neurons and multivibrators bring to mind similar structures with qubit-like properties.

Simulated Qubits

Recursive circuits as above have certain qubit-like properties, which, by the way, is why they are especially interesting. Advantages of qubit-like properties in a biological brain are, they can hold true and false simultaneously, releasing a logic value with a given probability when the circuit receives a read pulse. Moreover, a recursive circuit can be made to give a controlled toggle, which is a most useful type of neural operation.

Toggling works as follows: If probabilities are adjusted to give a *true* value with 100 % probability, a toggle would change (flip) this to *false* with 100 % probability. Similarly, false could flip to true. Controlled toggle circuits are termed controlled because they will flip only when other qubits in the system are true. They are quite useful for explaining mental arithmetic and other brain functions, so much so that they have been proposed to underlie gifted mental savants [3].

Recursive multivibrating neurons are easily converted into *simulated qubits*. Simulated qubits are discussed below and subsequently they will be compared with the qubits of quantum mechanics.

Recursive neurons as above, once triggered, sustain a sequence of pulses with similar amplitudes. Each pulse tends to have a brief width (τ) but the separation between pulses can vary; the frequency of pulses within a given neuron may range from a few hundred hertz to zero frequency, depending on neuronal delay parameters.

The general form of a simulated qubit is diagramed in Fig. 4.5. Unit A contains OR gate recursive circuitry as introduced above.

Note that the trigger to Unit A involves a weak synapse S1 to produce a single pulse, *Trigger(1)*. To ensure that the recycling pulse remains a single pulse, weak synapse S2 maintains a single pulse and serves to produce *Pulse*(*2*) which closes the loop. The waveform emerging on the right at point x will be sampled to establish either true or false. If the waveform is at a high level when sampled, the output will be true. If the waveform is at low level when sampled, the output will be false. This circuit provides true or false with a probability that depends on the frequency of multivibration.

To increase the computational possibilities, the relative phase of the output signal may be adjusted using *Delay*φ. This is referred to as *phase shifting* and its uses are explored in a later chapter.

The lowest frequency f_0 of pulses is defined to be state "zero" which can be related to false. The highest frequency f_1 is defined to be state "one" which can

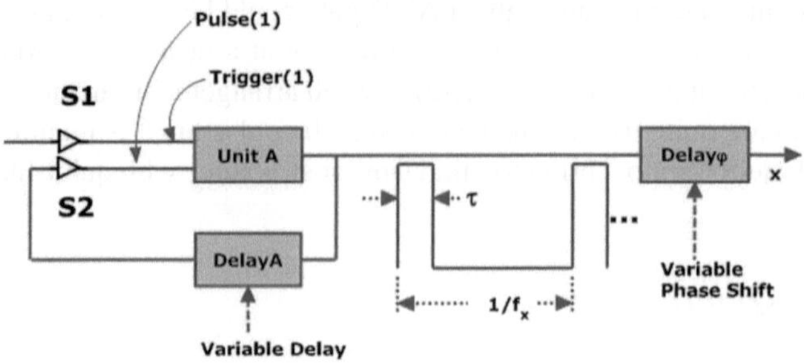

Fig. 4.5 Diagram of a simulated qubit

Fig. 4.6 Waveform with 50 % duty cycle

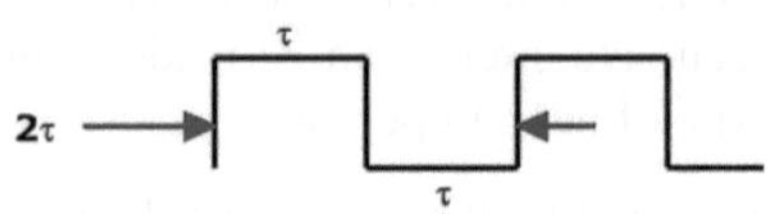

be related to true. It can be arranged so that the output is false with 100 % probability when at f_0 and true with 100 % probability when at f_1. Note that as a limit, $f_0 = 0$ is possible merely by opening the feedback loop through *Delay A*.

In general, frequency and duty cycle may be changed by adjusting *Delay A*. This is a neural multivibrator since pulses operate between a high and a low voltage level, and deliver a rectangular-shaped wave with a given *duty cycle*.

Duty cycle is a term that measures the separation of the same-sized pulses. Fifty percent duty cycle means that the pulses form a square wave such that pulse separation equals pulse width. The highest frequency f_1 for a recursive neuron is defined to have approximately 50 % duty cycle, shown in Fig. 4.6. This waveform is suggested to have the highest possible frequency because after a pulse, the neuron must recover sufficiently before another pulse is permitted.

Probability Formulations

Probabilities may be formulated by assuming a sampling pulse that approaches an impulse, one whose width is much less than τ. A lowest multivibrator frequency may be specified to be $f_0 \geq 0$ with a period of $T_o = 1/f_0$. The highest frequency may be specified to be f_1 with a period of 2τ. An intermediate frequency is specified to be f_x with a period of $T_x = 1/f_x$. Over several cycles, the ratio T_o/T_x equals the average number of cycles that f_x can deliver in time T_o. If f_x is a harmonic, say $n\,f_0$, with n integer, the ratio becomes $T_o/T_x = n$. A variable Γ is defined to be the average time per T_o that a waveform f_x is high:

$$\Gamma = \left(\frac{T_o}{T_x}\right)\tau. \tag{4.1}$$

The division T_o/T_x will be an integer only if f_x and f_0 are related harmonically. The fraction of time that a high occurs is

$$\frac{\Gamma}{T_o} = \frac{\tau}{T_x}. \tag{4.2}$$

This is proportional to probability for a true. The probability of a true is defined to be P which in percent may be written as

$$P = \tau f_x 100\ \%. \tag{4.3}$$

Using the above maximum duty cycle of 50 %, the ideal probability is 1/2 at f_1 and τ/T_o at f_0. The probability fraction of seeing a false is the complement of seeing a true:

$$P_{\text{False}} = [1 - \tau f_x]100\ \%. \tag{4.4}$$

Under this system, 100 % chance of seeing a false occurs only for $f_x = f_0 \rightarrow 0$.

Harmonics Assumption

Beginning with f_0, analysis of probabilities becomes easier if all frequencies are assumed to be synchronized, and if all frequencies are assumed to be related by integers, that is, harmonically related. This is a special case for the analysis, since in practice frequencies are not generally restricted to be harmonics of a fundamental.

Sampling Considerations

Multivibrator waveforms outputted at x are going to be sampled in order to capture a logic value. The sampling circuitry must provide a rest voltage if a low level is sampled, and a burst of pulses if a high level is sampled. Sampling should occur within a time frame T_o and sampling is usually just prior to when an output is

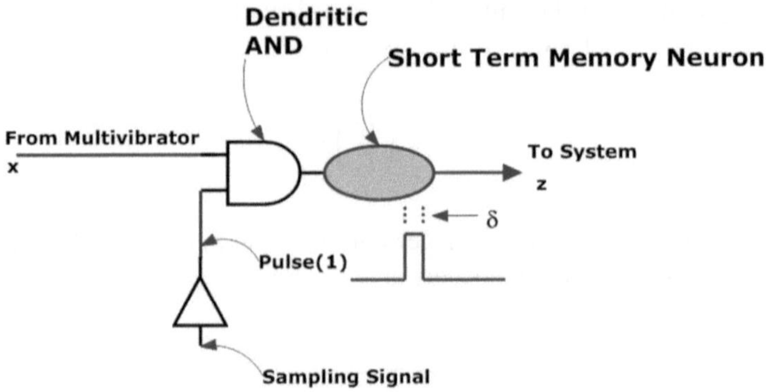

Fig. 4.7 Waveform sampling; z is either true (a pulse burst) or false (at rest)

available. Figure 4.7 illustrates how a sample might be taken. The weak synapse can produce a single pulse, *Pulse(1)* that will have a given pulse width δ, discussed below. If a single pulse happens to occur at the same time as a multivibrator pulse at the input of the AND gate, there will be a pulse at the output of the gate. This will trigger a long-term memory neuron to provide a regular burst at point z.

Two kinds of sampling may be identified: (1) Systematic and (2) Random.

Systematic Sampling

Systematic sampling is envisioned as being done assuming multivibrator frequency is either 0 or f_1; then either a true or a false will result with 100 % probability no matter when a sample is taken. A systematic sampling pulse might work for certain types of toggling circuits to make a simple true or false available in a timely way.

Random Sampling

In practice, pulses generated within the brain are likely to occur at convoluted times because of the differing locations of brain activities, and because such pulses are not synchronized with simulated qubit pulses. Such pulses are pseudorandom and often may substitute for random.

Truly random pulses might be generated in a receptor, one that is not part of a synapse, as suggested in Fig. 4.8. Such receptors are proposed as having a special structure so that they are sensitive only to especially energetic impinging ions, a random event, and so would generate a random dendritic pulse. Practical issues are that a truly random pulse might fail to occur when it is needed, or successive

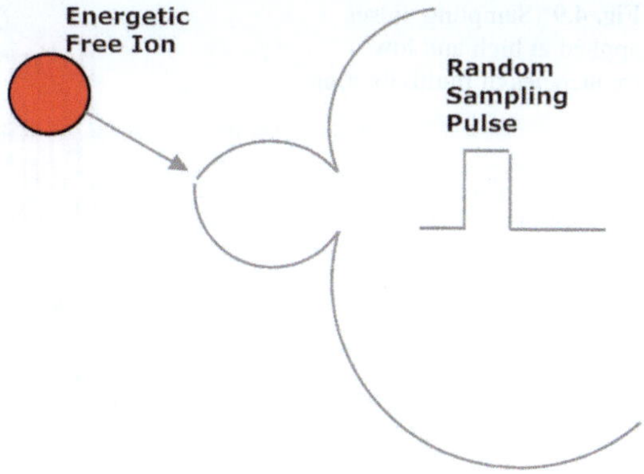

Fig. 4.8 Possible method to obtain a random sampling pulse

pulses might occur too close together. However, such issues usually make no difference biologically.

If it should happen that not even one sampling pulse occurs in a given cycle of T_o, more cycles of T_o are needed to assure that at least one random pulse has occurred. It may be noted that biological systems generally take the time they need; and in particular, they do not need to be fast. For example, a delay of a few cycles of T_o would not hurt the random masking of cues during a memory search, as done in the *Cue Editor* of a future chapter. Cue editing is done in the background and not in real time, so does not need to be fast.

As a note of little importance, but worth mentioning, when a multivibrator operates at f_0, the lowest frequency, the probability of a false is high, but the probability of a false does not quite 100 % unless f_0 can be made to be zero. To make probability of a false be 100 % for finite f_0, sampling could be synchronized with multivibration. Then sampling may be made to occur only between τ and T_o. When frequency is at f_0 this would guarantee a false with a probability of 100 %. Essentially the first pulse at the beginning of the period T_o is ignored.

Non-ideal Sampling

Non-ideal sampling is accomplished by a single pulse whose width is not zero but whose width δ as in Fig. 4.9. To assure that the highest frequency always gives an ideal true with 100 % probability, let $\delta = 2\tau$.

The probability of seeing a true as an output (after adjusted sampling) is

$$P \approx 2\tau f_x 100\,\%. \tag{4.5}$$

Note that a lower frequency for f_x lowers the probability of a true. The probability fraction of seeing a false is the complement of seeing a true:

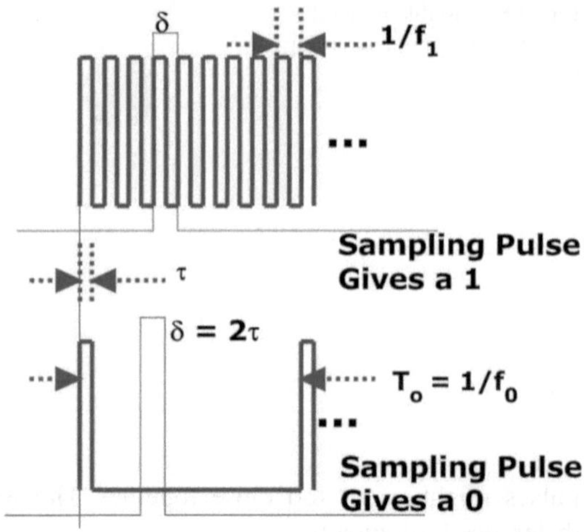

Fig. 4.9 Sampling pulses applied at high and low frequencies of multivibration

$$P_{\text{False}} \approx [1 - 2\tau f_x]100\ \%. \tag{4.6}$$

Using these equations, the probability fraction of seeing a true at f_0 is finite, although it is small (it is $2\tau/T_o$). This book generally assumes $f_0 \rightarrow 0$ where without doubt a 100 % chance of seeing a false occurs. Likewise at f_1 there is a 100 % chance of reading a true, assuming a sampling window whose width is $\delta = 2\tau$.

Frequency Control

Frequency is proposed to be controlled by circuit delay. Signal delay is regulated by the length of the circuit, membrane capacitance, and such current-charging parameters as the density of conductive pores (or ion channels) in an unmyelinated neural conductor and also by local ionic concentrations. Neurotransmitters may also have a role. Wide-ranging delay control is not used in this book; delay is modified only in situations in which delay must be increased very slightly. If needed, the proper combinations of parameters for delay control may be found by simulation. Figure 4.10 visualizes an idealized delay segment under control of a signal labeled y.

Variable frequency is a theoretical property of a simulated qubit but is not used very much in this book. However, controlled toggling, which is one of the main tasks of simulated qubits, is used a lot. To accomplish a toggle, the system simply needs to stop pulse cycling to achieve $f_0 \rightarrow 0$ and to trigger pulse cycling with minimum delay in the loop for the highest frequency f_1.

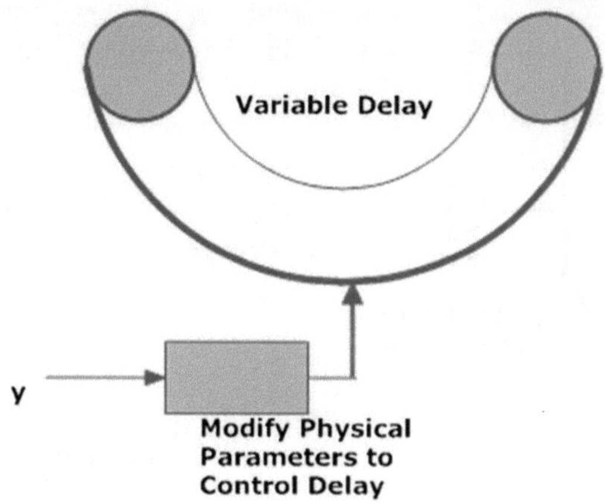

Fig. 4.10 Overview for delay control

Controlled Toggling

Controlled toggles are most convenient if they have a single input signal to trigger toggling. There are two ways to cause a multivibrator to toggle, as verified by simulation in the book's appendix. The first is termed *Excitory/inhibitory Stimulation*, the second is termed *AND/XOR Triggering*.

Excitory/Inhibitory Stimulation for Toggling

This method is perhaps the simplest in that a minimum of neural logic is required. The plan is to excite multivibration at one point in a loop and simultaneously to inhibit pulse propagation at a different (well chosen) point in the loop. Figure 4.11 illustrates the plan.

A *Single Pulse* is used to activate both synapses S11 and S12. Synapse S11 is a *fast excitory synapse* that triggers multivibration. S12 is a *fast inhibitory synapse* modeled with negative charge injection, with insignificant effect prior to multivibration, because there is no positive pulse in the vicinity of x3. To stop multivibration another *Single pulse to toggle* is applied to both synapses as in the figure. Multivibration stops because charge is drained from segments near S12, upsetting the balance necessary for pulse propagation. The time at which a *Single pulse to toggle* is applied apparently is not very important according to experimentation with simulations.

AND/XOR Toggling

This method of toggling requires a *Single pulse to toggle* to be applied at the correct time. Figure 4.12 shows the circuit. If there is no multivibration, a logic gate detects

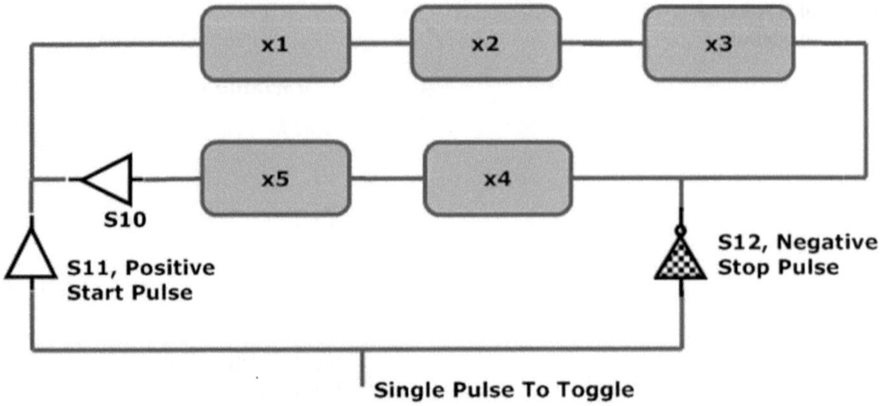

Fig. 4.11 Excitory/inhibitory toggling circuit

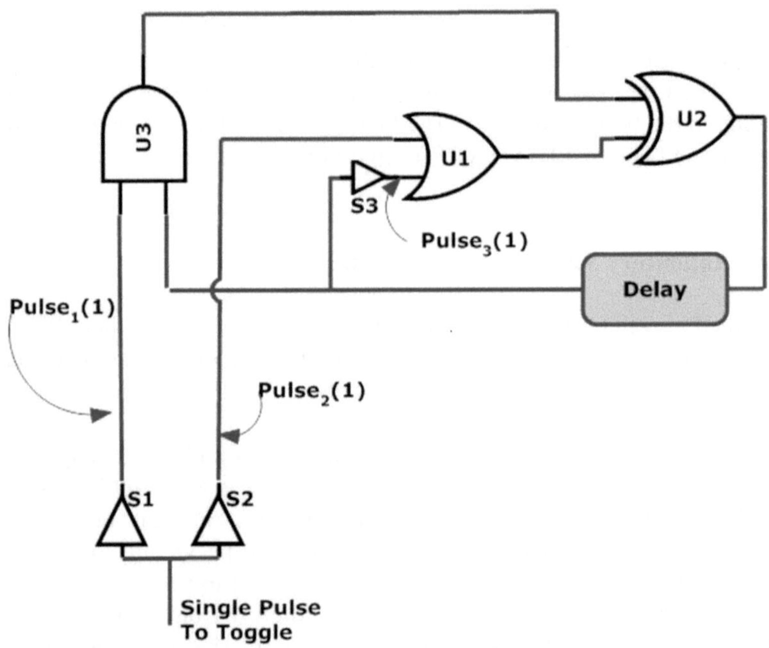

Fig. 4.12 AND/XOR toggle circuit

that there is no multivibration and permits normal triggering. Multivibration is triggered with S2. However, if multivibration is in progress, the gate U3 detects this and stops the cycling pulse via an XOR gate U2 whose output is made to be logic zero. Weak synapses S1, S2, and S3, by the way, are assumed to have insignificant delay. U1 and U3 are assumed to have identical delays so that pulses arrive simultaneously at U2.

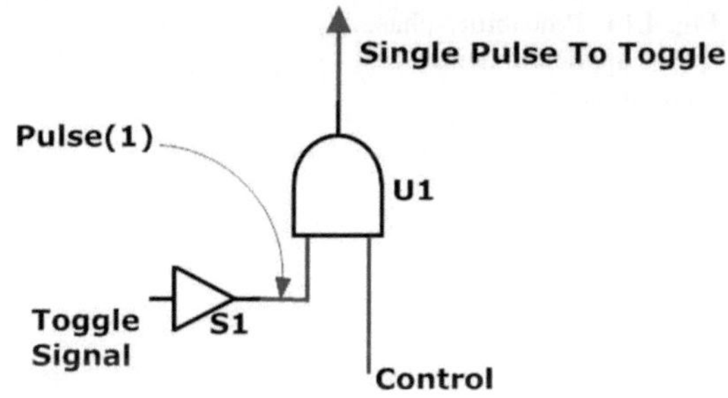

Fig. 4.13 Method of controlled toggling

To stop multivibration, the figure suggests that two simultaneous pulses open the AND gate U3, and therefore close the XOR gate U2, squelching multivibration. Details of a practical simulation of this are provided in an appendix to this book.

To stop multivibration, the *Single Pulse To Toggle* must occur within a given time frame in this circuit, that is, as a pulse is emerging from the delay segments. Communication to accomplish this might be done by synchronizing the state of the recursive circuit and the application of a toggle signal.

Controlled toggling means that there is a second signal that must be true if toggling is to occur. This may be implemented with a gate as in Fig. 4.13; the output of the gate produces the *Single Pulse To Toggle*.

The suggested circuits are not the only way of toggling, but these circuits have the virtue of simplicity. They are a form of long-term memory with computational possibilities and are very important to mental processing.

Sphere of Probability

A sphere of probability is an interesting way to visualize a single qubit; it shows the relevant variables (See Fig. 4.14). In a simulated qubit with an arbitrary mix of frequency and phase, an operating point can be visualized as a vector $\underline{\mathbf{a}}$ which will be somewhere on a sphere. The bold with underline refers to a simulated qubit vector, denoted as $\underline{\mathbf{a}}$. A physical qubit is denoted as $|\mathbf{a}>$.

Indirectly, the projection of vector $\underline{\mathbf{a}}$ on the z-axis is proportional to probability of a sampled true: If $\underline{\mathbf{a}}$ is on the axis and pointing up, there will be 100 % probability for a true, which is a Boolean 1; if $\underline{\mathbf{a}}$ is on the axis and pointing down, there will be 0 % probability of a true or 100 % probability for a false, which is Boolean 0. The figure shows approximately a 50–50 mix of the two states, 0 and 1. A change of phase φ, or *delay*φ in the simulated qubit, causes the vector to rotate about the z-axis, but this has no effect on a projection on the z-axis or on the resultant probability.

Consider two multivibrators each with 50 % chances for true or false. When read with independent random pulses there is a 25 % chance of ending up with one of $\underline{00}$, $\underline{01}$, $\underline{10}$, $\underline{11}$. Four states are present. For n multivibrators, 2^n possible states

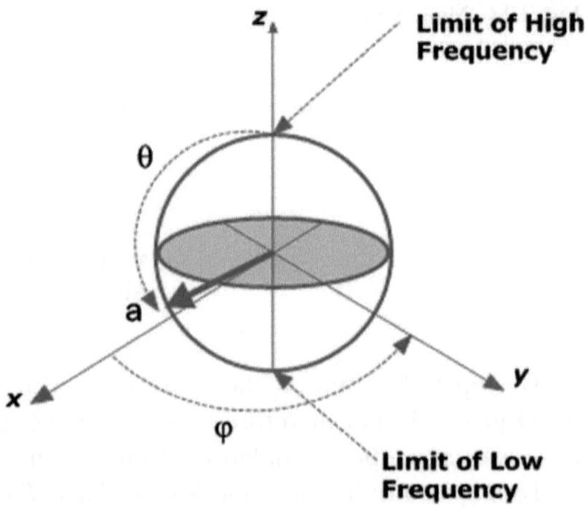

Fig. 4.14 Probability-phase sphere applied to neuro-multivibrators

are possible, each with a certain probability. Prior to sampling, the vectors that represent these outcomes are known in the abstract as *probability space*.

The author places **0** as being a lower frequency on the bottom of a sphere, since low usually means closer to bottom (usually in quantum physics, $|\mathbf{0}>$ is at the top).

Physical Qubits

Analogy to Physical Qubits

A physical qubit is an entity in a quantum system with appropriate quantum properties, for instance, a rotating electron with two quantum states, spin up and spin down. These may be thought of as rotation counterclockwise and clockwise. Spin up and spin down defines two basis states $|\mathbf{0}>, |\mathbf{1}>$ which may be expressed as vectors:

$$|\mathbf{0}> = \begin{bmatrix} 1 \\ 0 \end{bmatrix}; \quad |\mathbf{1}> = \begin{bmatrix} 0 \\ 1 \end{bmatrix} \tag{4.7}$$

These can be written as $|\mathbf{0}> = [1\ 0]'$ and $|\mathbf{1}> = [0\ 1]'$ (the prime denotes a transpose and is a way to express a vertical vector on a horizontal line). Qubit vectors may combine linearly. A general expression for a single qubit state vector is a mathematically linear combination of the two possible states:

$$|\psi> = \alpha|\mathbf{0}> + \beta|\mathbf{1}>. \tag{4.8}$$

This implies the surprising situation that a qubit can be in two logical states at once. In general α and β are complex numbers. According to the quantum model $|\alpha|^2$

is the probability of $|\mathbf{0}>$; $|\beta|^2$ is the probability of $|\mathbf{1}>$. To properly conserve probability, a mathematical constraint known as the normalization condition must apply to qubits:

$$|\alpha^2| + |\beta^2| = 1. \tag{4.9}$$

This is an equation for a sphere and suggests a point on the surface of a sphere [4, 5]. For equal proportions of true and false probabilities, a state vector could be denoted as $|\mathbf{a}>= \eta[1\ 1]'$, where $\eta = 1/\sqrt{2}$ achieves vector normalization, that is, $(1/\sqrt{2})^2 + (1/\sqrt{2})^2 = 1$. This means 50 % chance of observing a $|\mathbf{0}>$ and 50 % chance of observing a $|\mathbf{1}>$.

The probability sphere for a given qubit portrays only the relative phase of α relative to β. In general both α and β may have their own independent phases, but often it is the phase difference that is important to quantum calculations.

Note that $|\mathbf{a}>$ is thought of as rotating through an angle θ within the x–z plane about the y-axis. Relative phase shift involves another angle φ in the x–y plane about the z-axis. The result is that $|\mathbf{a}>$ can have any direction. In particular, negative signs are permitted, for instance, either $+|\psi>$ *or* $-|\psi>$ are possible. This is considered a phase reversal. Note that the sphere can be misleading since it does not show the phase for an ideal $|\mathbf{0}>$ or $|\mathbf{1}>$; but both $+|\mathbf{1}>$ and $-|\mathbf{1}>$ would read out as a one with 100 % because phase does not affect the probability.

Suppose now that two qubits are available in a quantum mechanical system. Thus there are four possible states. These states can be expressed using a *direct product*, to be taken as below. A direct product is sometimes indicted with $\otimes$, for example $|\mathbf{a}> \otimes |\mathbf{b}>$.

$$|\psi>= |\mathbf{a}> \otimes |\mathbf{b}>= [a_1|\mathbf{b}>\ a_2|\mathbf{b}>]'$$

$$|\mathbf{a}>= \begin{bmatrix} a_1 \\ a_2 \end{bmatrix}, \quad |\mathbf{b}>= \begin{bmatrix} b_1 \\ b_2 \end{bmatrix} \tag{4.10}$$

In this case a direct product gives the state vector as

$$|\psi>= \begin{bmatrix} a_1 b_1 \\ a_1 b_2 \\ a_2 b_1 \\ a_2 b_2 \end{bmatrix} \tag{4.11}$$

For illustration, assume each qubit is prepared to have probability levels of 50 %. Then $|\mathbf{a}>= \eta[1\ 1]'$ and $|\mathbf{b}>= \eta[1\ 1]'$. Upon readout, there is a 25 % chance of any given combination $|\mathbf{00}>$, $|\mathbf{01}>$, $|\mathbf{10}>$, $|\mathbf{11}>$. This may be expressed as

$$|\psi>= \eta^2 \begin{bmatrix} 1 \\ 1 \\ 1 \\ 1 \end{bmatrix}. \tag{4.12}$$

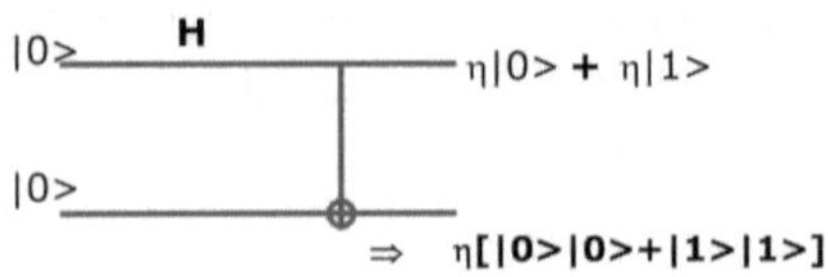

Fig. 4.15 "Wiring" diagram example

The direct product symbol $\otimes$ is usually understood from the context of a discussion of state vectors and not written. So $|\mathbf{a}>\ |\mathbf{b}>$ means $|\mathbf{a}>\otimes|\mathbf{b}>$. Subject to the normalization conditions, $|\psi>$ may be filled with fractions of differing values such that: $0 \le a_1,\ a_2,\ b_1,\ b_2 \le 1$. Note again that two qubits resulted in four states. Generally n qubits have 2^n states.

Classical Simulated Qubits Versus Quantum Qubits

The electron, for instance, can exist partly with spin up and partly with spin down. In contrast, an intermediate multivibrator frequency is not a combination of low frequency f_0 and high frequency f_1. The probability mix is only "simulated." This is a major difference.

Other lesser differences are: There is only one phase variable in the above version of a simulated qubit whereas a physical qubit has two phase variables, one for $|\mathbf{0}>$ and one for $|\mathbf{1}>$. That is, for α and β. The above simulated qubit has only one phase variable. Clever design can bring in another phase variable for simulated qubits, as in another chapter, but it is unnecessary for most neural situations.

If the waveforms from multiple multivibrators are synchronized (all start at the same moment), and are harmonically related, probabilities take on discrete values, unlike quantum probabilities, since the available frequencies are assumed to go up as integers. Thus, if $f_0 = 1$ Hz and f_x is an integer up to a few hundred, probability from the above equations takes on discrete values.

Another difference is that in classical simulation entanglement is impossible.

Entanglement

An important feature of physical qubits is that they can be entangled. Consider two qubits $|\mathbf{a}>$, $|\mathbf{b}>$ initialized to $|\mathbf{0}>$ in a quantum mechanical system. The entanglement process is diagrammed in Fig. 4.15. The following theoretical procedure may be applied. Step (1) Qubit $|\mathbf{a}>$ may be prepared to be $|\mathbf{a}>= \eta[1\ 1]', \eta = 1/\sqrt{2}$. Step (2) A controlled NOT may be applied such that if $|\mathbf{a}>$ is $|\mathbf{0}>$, nothing happens; but if $|\mathbf{a}>$ is $|\mathbf{1}>$, then $|\mathbf{b}>$ is flipped (inverted) from $|\mathbf{0}>$ to $|\mathbf{1}>$, for that fraction of $|\mathbf{a}>$ that is $|\mathbf{1}>$. The result will be two entangled qubits with an equation of the form

$$|\psi>= \eta\big[|\mathbf{0}>\ |\mathbf{0}> + |\mathbf{1}>\ |\mathbf{1}>\big]' \tag{4.13}$$

This can be expanded to the following entangled form:

$$|\psi>=\eta\begin{bmatrix}1\\0\\0\\1\end{bmatrix}. \tag{4.14}$$

Note that the transform **H**, given below, creates η [1 1]′, the top line. The controlled NOT is controlled by the 1 component of the top line and causes a NOT in the bottom line.

Entanglement could be interpreted to mean that the resultant $|\psi>$ is not a direct product of two qubits (without using addition somewhere). In contrast, a non-entangled state vector can be factored into a product involving individual qubits, for example, the following state vector is non-entangled since it can be factored:

$$|\psi>=\eta^2\begin{bmatrix}1\\1\\1\\1\end{bmatrix}=\mathbf{H}|\mathbf{0}>\otimes\mathbf{H}|\mathbf{0}>, \tag{4.15}$$

where

$$\mathbf{H}=\eta\begin{bmatrix}1 & 1\\1 & -1\end{bmatrix} \tag{4.16}$$

H is a transformation that converts $|\mathbf{0}>=[1\ 0]'$ into $\eta[1\ 1]'$ and the reverse, $\eta[1\ 1]'$ into $[1\ 0]'$. This reversible operation is assumed to be available physically.

If two entangled qubits are physically separated by a distance, while remaining in a given quantum system, they would continue to be entangled. Say qubit $|\mathbf{a}>$ is observed to be $|\mathbf{0}>$. Then at a later time when qubit $|\mathbf{b}>$ is far removed, and when observed, it will also be a $|\mathbf{0}>$. Similarly if a $|\mathbf{1}>$ is read for $|\mathbf{a}>$, then $|\mathbf{b}>$ has to be a $|\mathbf{1}>$. This feature is termed *teleportation* and theoretically occurs without physical connections. For this to work, the qubits must not lose coherence because of thermal agitation. It would be very exciting to discover someday that teleportation occurs within neurons causing hidden communications.

Simulated qubits, although potentially useful in a biological environment, cannot achieve teleportation. A connecting interneuron is necessary. Simulated qubits cannot achieve teleportation, but they can be synchronized, that is, they can be made to start at the same time. And they can be made to be sampled at the same time using a common sample pulse. This is suggested in Fig. 4.16, showing that similar frequencies give exactly the same sampled logic outputs. So in a sense, the state of one simulated qubit determines the state of the other.

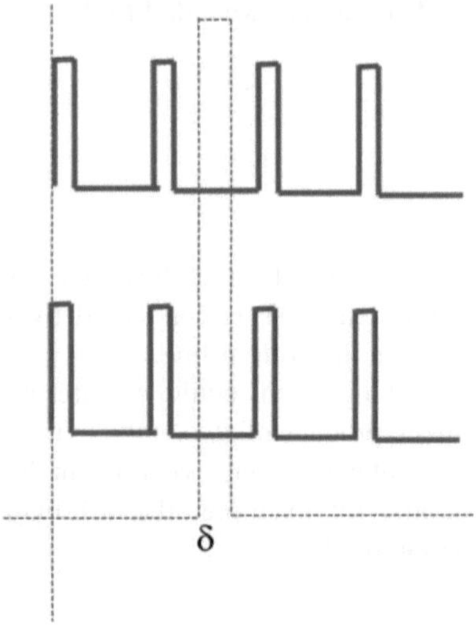

Fig. 4.16 Synchronized simulated qubit forcing identical sampled logic outputs

Conclusions

This chapter has introduced brain memory elements and illustrated them with physical circuits. These circuits employ circuit elements as necessary to explain complex neural operations. Concepts using LTP partly explain biological long-term memory, but there is also the possibility of recursive neurons for memory. Recursive multivibrating neurons are compatible with known features of human memorization, such as instant, photographic memory, and indefinitely long-lasting memory. A combination of LTP and multivibration is proposed.

Recursive neurons with qubit properties are termed *simulated qubits*; they simulate some, but not all, features of physical qubits. Of interest is that a single neuron may simultaneously hold true and false with a given probability. Probabilistic logic is of interest scientifically and also biologically, such as for cue editing.

Simulated and physical qubits are compared in this chapter. It was noted that simulated qubits lack the possibility of teleportation, whereas real qubits, if entangled and physically separated, continue to be in communications without a (known) physical medium. To achieve teleportation, it is necessary to maintain an undisturbed quantum system, making it difficult in practice to separate qubits appreciably.

Qubits within brain cells at the molecular level would increase mental abilities by orders of magnitude, so this possibility is too important to ignore. Simulated qubits might also be possible at the molecular level within a brain, without the difficulty of maintaining a quantum system for significant duration in the face of thermal agitation.

What is most important about simulated qubits is that they can produce controlled toggles, as suggested by circuitry provided in this chapter, and simulations in the appendix. Controlled toggling is important to arithmetic and may be important to subconscious mental arithmetic as demonstrated by gifted savants. A later chapter explores possibilities for a specific type of toggled arithmetic necessary for the prioritization of memory returns.

References

1. Kanerva P (1988) Sparse distributed memory. MIT Press, Cambridge, MA
2. Bliss T, Collingridge GL, Morris RGM (2004) Long-term potentiation: enhancing neuroscience for 30 years. Oxford University Press, Oxford
3. Burger JR (2011) Qubits underlie gifted savants. NeuroQuantology 9:351–360
4. Nielson MA, Chuang IL (2000) Quantum computation and quantum information, Cambridge series on information and the natural sciences. Cambridge University Press, Cambridge [Paperback]
5. Pittenger AO (1999) An introduction to quantum computing algorithms. Birkhauser, Boston

Self-Study Exercises

(If you desire to keep a record of your efforts, please show a couple of logical steps leading to the answers.)

Long-Term Potentiation

1. Assume a receptor membrane pulse (instigated by sodium and other ions) to have a steady value of 50 μA/cm^2 and assume that the pulse is 200 μs wide. Assume that receptor membrane capacitance is 1 μF/cm^2.

   (a) Calculate the change in membrane voltage. (ANS. 10 mV)
   (b) Initially membrane voltage is at rest at −70 mV. Assume that it can be triggered to provide a regular neural pulse if voltage exceeds −55 mV. Does it trigger? (ANS. No because the voltage raises only to −60 mV)
   (c) Assume LTP to a level of 10 mV above the rest voltage of −70 mV. Does the membrane produce a neural pulse? (ANS. Yes because with LTP it is resting at −60 mV. An additional 10 mV causes voltage to exceed −55 mV, the threshold for triggering)

Multivibratation

2. Assume a pulse width of 1 ms, OR gate delay of 2 ms, and a feedback loop delay of 10 ms.

(a) What is the frequency of multivibration? (ANS. 83.3 Hz)
(b) Assume an ideal pulse with a required minimum duty cycle of 50 %. What would be the highest frequency of multivibration? (ANS. 500 Hz).

Hybrid Memory Element

3. Consider how to stop multivibration:

   (a) Estimate values for Delay3 and Delay 2 to stop multivibration, assuming X = 10 ms delay in each gate component. (HINT: Delay3 ≈ Delay2 + 2X; it remains to choose Delay2 to give about ten pulses)
   (b) Redesign the hybrid element using an AND gate.

Simulated Qubits

4. Assume a waveform with a pulse width of 1 ms and a $f_0 = 1$ Hz.

   (a) What is the ideal probability that the waveform is true? Assume that the sampling pulse is an impulse with zero width. (ANS. 0.001 or 0.1 % probability)
   (b) Continue to use the reference duration of $1/f_0$ but assume that the actual frequency of multivibration is $f_x = 100$ Hz. What is ideal probability? (ANS. 10 %)
   (c) Continue to use the reference duration of $1/f_0$ but assume that the actual frequency of multivibration is $f_x = 0.5$ kHz. What is ideal probability? (ANS. 0.5 or 50 %)

5. A multivibrator produces a pulse width τ. Predict the probability of a true occurring under the following conditions:

   (a) Frequency is at its maximum of f_1 and an ideal sampling pulse whose width is much less than τ is available. (ANS. Duty cycle is 50 %, average probability of a true is 50 %)
   (b) Frequency is at its maximum of f_1 and a sampling pulse whose width is 2τ is available. (ANS. Duty cycle is 50 %, average probability of a true is 100 %)
   (c) An ideal sampling pulse (zero width) and frequency is reduced by 50 %. (ANS. Duty cycle is 25 %, average probability of a true is 25 %)
   (d) An sampling pulse whose width is 2τ is available and frequency is reduced by 50 %. (ANS. Duty cycle is 25 %, average probability of a true 50 %)

6. Assume a waveform with a pulse width of 1 ms and a $f_0 = 1$ Hz. Assuming a sampling pulse that is 2 ms wide and occurs in a reference duration of $1/f_0 = 1$ s,

and will register a true if it happens to fully contain an equivalent of a full multivibration pulse.

(a) What is the probability of sampling a true? A false? (ANS. 0.1 %, 99.9 %)
(b) Repeat the above, only assume sampling is synchronized with f_0 to begin at 1 ms and to end at 1 s. What is the probability of a false at f_0? (ANS. 0)

7. Derive, with reference to sketched waveforms, that $P = \tau f_x$ 100 %.

Toggle Circuits Using Simulated Qubits

8. State the differences between a toggle and a data latch. (HINT: a data latch holds the truth value applied to it until it is overwritten or cleared, after which the original truth value is lost forever; by keeping track of the number of toggles, past truth values in a toggle are never lost)
9. Draw the given circuit for Excitory/inhibitory toggling. Explain how it is started and how it is stopped.

   (a) Why does it start?
   (b) Why does it stop?

10. Draw the circuit for the NAND/XOR toggling and explain how it is started and how it is stopped.

    (a) Why does it start?
    (b) Why does it stop?

Qubit Sphere

11. Approximately locate

    (a) The vector with zero phase and 75 % probability for a true.
    (b) The vector with 180 ° phase and 75 % probability for a true.
    (c) Why does phase have no effect on the measured probability?

Qubits

12. Calculate the 16 elements of the following direct product in symbolic form: (HINT. First element would be *ae*, second element would be *af*. . .)

$$\begin{bmatrix} a \\ b \\ c \\ d \end{bmatrix} \otimes \begin{bmatrix} e \\ f \\ g \\ h \end{bmatrix} = ?$$

13. Calculate the four elements of the following matrix combination in symbolic form: (HINT. Element (1,1) would be $ae + bg + w$)

$$\begin{bmatrix} a & b \\ c & d \end{bmatrix} \begin{bmatrix} e & f \\ g & h \end{bmatrix} + \begin{bmatrix} w & x \\ y & z \end{bmatrix} = ?$$

14. Prove that the 2×2 Hadamard matrix $\mathbf{H}$ is its own inverse $\mathbf{H}^{-1}$. (HINT: Show that $\mathbf{H}\,\mathbf{H}^{-1} = \mathbf{1}$, a diagonal matrix with ones on the diagonal)

$$\mathbf{H} = \begin{bmatrix} 1 & 1 \\ 1 & -1 \end{bmatrix}$$

15. Provide a normalized state vector of dimension four with 70 % chance for the first given combination to be observed, and 10 % chance for the remaining three combinations to be observed. (ANS. [0.8366 0.316 0.316 0.316]′)
16. Assume an entangled state vector with two qubits of the form $|\psi >= \eta(|\mathbf{0} > |\mathbf{0} > + |\mathbf{1} > |\mathbf{1} >)'$

    (a) When this vector is observed, what is the probability of a zero in the first qubit? (ANS. 50 %)
    (b) If a zero is observed in the first qubit, what is the probability of a zero in the second qubit? (ANS. 100 %)
    (c) Assume the qubits are physically separated and that the second qubit is taken to the other side of the world. Assume an observation of the first qubit indicated that it was a one. What is the probability of seeing a one for the second qubit when it is observed? (ANS. 100 %)
    (d) What are the conditions on these physically separated qubits? (ANS. They must continue to be in a coherent quantum mechanical system and thus continue to be entangled)

Differences Between Simulated Qubits and Physical Qubits

17. List all differences between simulated and physical qubits.
18. Show with sketches of waveforms, that synchronized multivibrator qubits at the same frequency will provide identical outputs if sampled with synchronized pulses.
19. Simulated qubits applied as controlled toggles may operate as an array in which one qubit, if its output is true, causes another to toggle. This requires a bus.

    (a) Design a primordial processor using four controlled toggle circuits. You may use numbered mechanical switches as required to open and close circuits to accomplish toggles.
    (b) Initialize 16 toggles to zero. Write a procedure to determine whether or not *one* of the toggles was secretly changed to "1" behind your back. (HINT: Include a 17th toggle initialized to zero to serve as a "flag." It is successively connected to each output and is toggled only if it sees a "1.")
    (c) Under what conditions would the flag fail. (HINT: If an even number of 1s occur)

Chapter 5
Outline of a Cue Editor

Introduction

This chapter introduces cue editing. Usually what passes through conscious short-term memory (STM) is a result of sensory inputs, visual images, and sounds, for instance. Conscious STM in turn serves to provide cues that are required to produce associative returns from subconscious long-term memory [1–3]. These returns, if they have priority, will replace the current contents of conscious STM. A *cue editor* is envisioned as kicking in when cues are flawed, taking important attributes from conscious STM and manipulating them to help guarantee associative returns from subconscious long-term memory.

Cue editing is necessary because STM attributes quickly fade, soon to be replaced by other thoughts. Under fading conditions, attributes can be incomplete or inconsistent. For instance, one image might mix with another, providing cues that will not associate with anything. For confused cues a cue editor is essential.

A similar situation occurs for fleeting recalls from subconscious memory, frequently a source of cues for new recalls. Important attributes may be missing, or past images may leave a residue of unrelated cues. Cue editing is necessary.

The specifications for a cue editor follow from what it must do, which is to assure associations by modifying the available cues in the sense of removing cues one or two at a time, and if this fails, restoring them, and removing different ones. Cues are modified very slightly at first, and if this fails, a larger number of them are modified at a time. Without a cue editor, significant dead time in consciousness would frequently result, with unfortunate mental blocks; and this would significantly reduce the survivability of humans.

A cue editor also supports subliminal search. To recap what this means, everyone has experienced a memory block. But unknown to a person, a subliminal search proceeds for what is trying to be remembered. Cues keep cycling subconsciously, working to achieve a correct recall. When a person least expects it, the correct memory will click into consciousness. This is an indication that the brain has worked in the background, changing cues until they achieve useful recalls.

J.R. Burger, *Brain Theory From A Circuits And Systems Perspective*, Springer Series in Cognitive and Neural Systems 6, DOI 10.1007/978-1-4614-6412-9_5,
© Springer Science+Business Media New York 2013

A reader must keep in mind that a cue editor does not judge relevance of cues, this being the purpose of a *recall referee* to be described in subsequent chapters.

Brain System Environment

There is no fixed schematic of a cue editor, just like there is no fixed schematic for a brain as far as we know; probably each brain differs. But there seems to be a general plan which may be implemented in various ways. The electrical and logical designs offered below demonstrate a particular implementation plan for a cue editor.

When modeling neural circuits, a full assortment of dendritic logic and enabled (soma) logic are available. Also available are regular synapses to excite pulse bursts, and weak synapses with an ability to excite a single pulse for precise timing. Also of interest are specialized STM neurons to provide bursts of specified lengths. All of the above circuit elements are involved, although in the interests of readability and in keeping with the spirit of an overview, generic logic symbols are used as much as possible. For similar reasons timing diagrams and requirements are not emphasized below, even though such considerations are important in a practical implementation.

Cue Editor Overview

Cues are assumed to be provided by images imposed by the body senses, and also images from long-term memory. When a new image appears in conscious STM, its attributes linger for a while, usually a brief while until updated. During this brief duration cues taken from attributes are available for memory searches.

Proposed Architecture

Figure 5.1 suggests a possible architecture for a cue editor. There is a neural bus with *K* axons to carry a maximum of *K* cues. The bus symbol /*K* means a bus with *K* conductive paths. There is a control line labeled *New*(*i*), where $i = 1, 2, 3, \ldots D$; "*i*" is an index mainly for explanatory purposes. *New*(*1*) goes true (emits a signal) to signal a new set of cues, perhaps from the first image after awakening. Then comes *New*(2) and so on throughout the day.

The logic will be such that immediately after *New*(*1*) goes true, there is a search of long-term memory for an association with the given cues. Most often there is a match, or a *Hit*, after which the system waits for *New*(2), and the next set of cues.

But if no exact matches are found, a *NoHit* signal is emitted from long-term memory. In this event cues are loaded into a register *R1* assumed cleared and ready

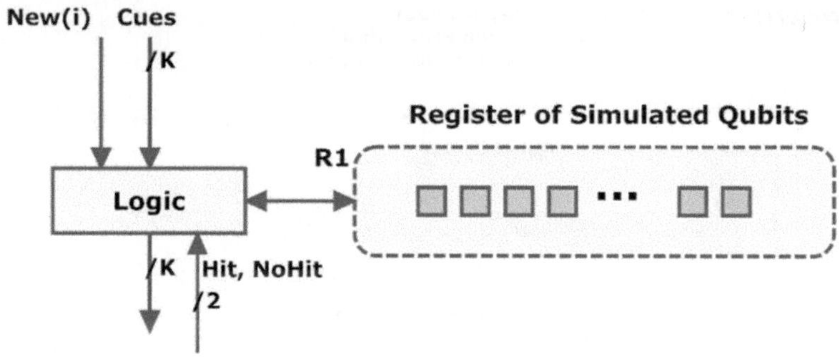

Fig. 5.1 Cue editor proposed architecture

for attributes. *R1* in the figure consists of an array of simulated qubits with memory and processing ability, to be discussed later. These simulated qubits will be made to randomly remove cues in an attempt to obtain returns from long-term memory. If returns still do not occur, all cues are restored in *R1* and another random selection is removed. The search process continues this way until interrupted by a new, higher priority set of cues taken from a new image in conscious STM.

Another Subliminal R1 search is not implemented if *New*(2) happens to go true, for example, in which case a new set of cues are processed from STM. New cues have higher priority over past inconsistent cues.

Should it happen that *New*(2) also fails to produce a return, the logic will be such that cues for *New*(2) are shuffled into *R1*, whose cues associated with *New*(*1*) are erased and lost forever in this specification. Additional registers, *R2* for example, are theoretically possible, but would unnecessarily complicate the presentation of this chapter.

Cue Editor Logic

This section is going to provide an outline of the logic for a cue editor, but not so exhaustive as to be a rigorous specification. The goal is to demonstrate the viability of the above architecture. Figure 5.2 shows the top level. When a new image first appears in conscious STM, *New*(*1*) goes true. Axons working in parallel are enabled by *New*(*1*) via an array of AND gates labeled U1, that is to say, Unit 1; this applies available cues to associative long-term memory via U3.

R1 initially is cleared so that $Flag = 0$. This shuts down gate U8, forcing *CueMods*(*1*) to be logic 0, which in turn disables U2, U7, and U9. This means that *R1* cannot interrupt the current search until the results of the search become apparent.

If *Hit*(*1*) is returned with a true, nothing more happens in *R1*. This means a match was found, and the system simply waits for the next *New*(2). But if *NoHit*(*1*) is returned with a true, cue editing will begin, because gate U10 activates *Load*(*1*).

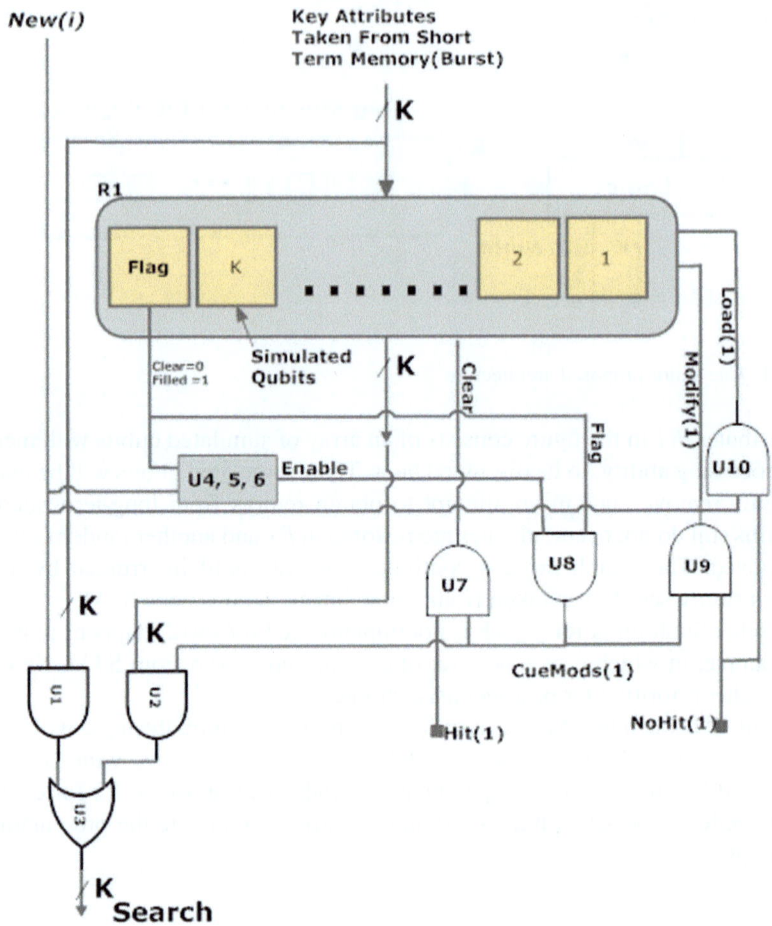

Fig. 5.2 Proposed subcircuit for cue editing

U10 is activated because *New(1)* has prepared it for what occurs next. A *NoHit(1)* signal causes the existing inconsistent cues, shown coming down from the top, to be loaded into *R1*. This is a register of simulated qubits that are configured to begin their work as simple registers, that is, respective qubits go true (frequency goes to f_1) if their respective attribute is present, or false otherwise (frequency in this case goes to 0).

If *NoHit(1)* occurs, *R1* undergoes *Load(1)* and also makes *Flag* go true. It is now assumed that *New(2)* $= 0$, and that *Enable* is true. The block U4, 5, 6 is going to be discussed below. Thus U8 is enabled, and *CueMods(1)* goes true. This enables U2, U7, and U9, and makes possible a *Modify(1)* and also a *Clear* in the event that *Hit(1)* occurs.

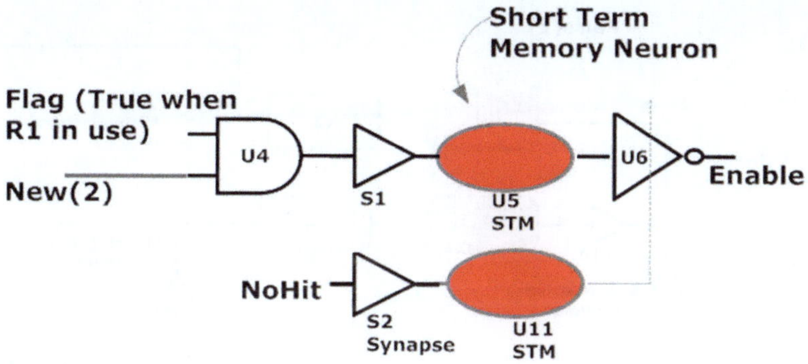

Fig. 5.3 Block U4, 5, 6

Soon there will be a *double-check search* as cues are applied via U2 and U3. If there is an unexpected match, *R1* is cleared via U7 and the system is taken back to its initial condition. But if *NoHit(1)* goes true again, the cues are now modified via U9 in that the probabilities for each true attribute are taken below 100 % as described below. This means their multivibrators have $f_x < f_1$. Another search occurs asynchronously with the modified cues. The process of modification and searching continues intermittently until *Hit(1)* occurs and a *Clear* results, or until a new set of conflicting cues is sent to *R1*.

Block U4, U5, U6

Block “U4, 5, 6” generates an *Enable* as now described (see Fig. 5.3). This is used to disable the registers for a specified time, as necessary when a higher priority search is called for, meaning a new image in conscious STM and *New*(2) = *True*. When *Flag* = *True*, and *New*(2) = *True* U4 is activated. The output of U4 goes true and activates a synapse S1 that forces the output of U5 to go true for several milliseconds. U5 is a STM neuron. The *NoHit* signal enables U11, another STM, creating an environment for an inverter U6. The inverter is assumed to use inhibitory synapses as described in a previous chapter.

The output of inverter U6 subsequently goes false, so *Enable* is false. This prevents the output of U8, and *CueMods*(*1*), from going true. While *CueMods*(*1*) is false, it is not possible to search with unsatisfied cues that have been transferred to *R1*. Rather, the most recent cues associated with New(2) are passed through and processed first. The duration of disabling must exceed the time to conduct a memory search, a few tens of milliseconds.

If the search enabled by *New*(2) finds a match, the background search associated with *Flag* will start up again. But if *NoHit*(2) happens to go true for this new search, conflicting cues for this new search are automatically loaded into *R1*, erasing the first set of cues held there.

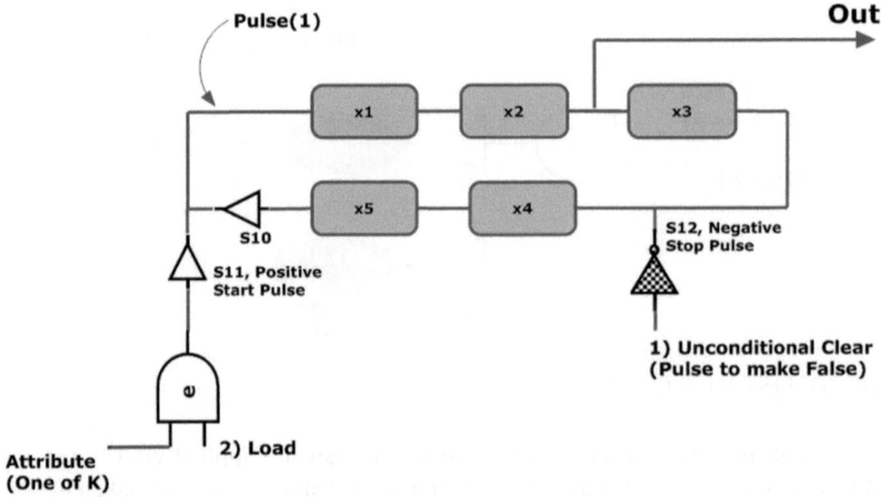

Fig. 5.4 Initializing a toggle circuit to false output

The most recent *New*(*i*) takes higher priority and will block background searching, which will resume its actions only when there are no new sets of cues coming down from short-term memory. Its actions are resumed as soon as the short-term memory U5 is timed out after a few tens of milliseconds. We note again that *R1* is not expected to be needed on a regular basis, but on those critical occasions when the cues are in conflict, it is essential.

Operations in *R1*

Initially the inconsistent cues are loaded into *R1* and the *Flag* qubit is set true. At an early point *New*(*1*) returns to false, so U10 prevents reloading *R1* until proper processing takes place. *Enable* moves to true, so the output of U8, *CueMods*(*1*), goes true. Via U9, when *NoHit*(*1*) is true, a *Modify*(*1*) command goes to *R1*. This will cause a change in the probabilities of the given cues. Subsequently another search occurs when the modified cues flow to long-term memory through U2 and U3.

Next we explain operations in *R1*, the simulated qubit circuits.

Simulated Qubit Operations

Simulated qubits are used in the form of toggle circuits as introduced in a previous chapter. To begin, all toggles are reset, or cleared to false. The easiest way to do this is to use inhibitory synapses as in Fig. 5.4. There are two steps for loading *R1*.

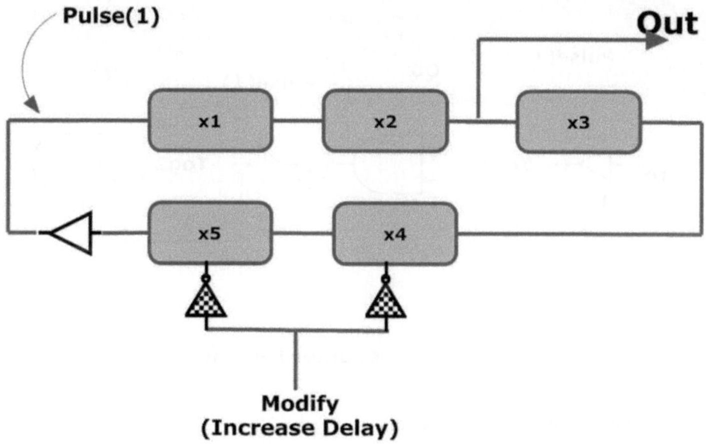

Fig. 5.5 Modification of cue probabilities

1. Toggles are pulsed with an *Unconditional Clear* to reset the output to false. This nonreversible operation is equivalent to setting the output to false if it was originally true, and leaving it false if it was false.
2. A *Load* applies true attributes, when attributes are true, that serve to trigger multivibration. This is accomplished for each multivibrator in *R1*. This effectively places all cues into an array of simulated qubits. Note that the *Flag* qubit must also be set as a result of a load command.

Modifying Cue Probabilities

For those cues that are true, a modification is desired to lower their probability of being true. This is equivalent to lowering their frequency of vibration. Lowering frequency involves increasing delay very slightly, assumed to be possible with special synapses as suggested in Fig. 5.5.

It is assumed that delay can be increased slightly with inhibitory-like synapses that serve to reduce the rate of charging into membrane capacitance. If it takes longer to charge, signals propagate slower. Increasing delay is equivalent to switching in a short length of interneuron to increase delay. Increasing delay causes the frequency of multivibration to decrease and results in a chance that true cues will be sampled to be false, which is the desired effect.

Note that frequency f_x may be reduced smoothly. In this case it is not restricted to be a harmonic of some lower frequency f_0.

Upon sampling, which is not shown, cues are thus removed at random to increase the chance of a match in long-term memory. Each time the simulated qubits are sampled again, all cues are effectively restored, and different cues are removed. This system will not add cues or attributes that were not present initially.

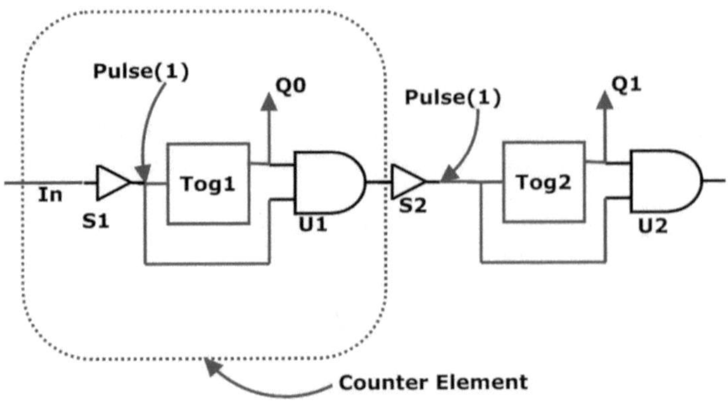

Fig. 5.6 Pulse counter using toggle elements

The *ModCues* signal is reapplied after each failure to find a match, that is, for each *NoHit* signal. *ModCues*(i) is assumed to lower the frequency of multivibration very slightly each time it is applied $i = 1, 2, 3, \ldots$. Consequently, as successive modifications occur there will be an ever-decreasing number of cues, and an ever-increasing chance of a match in long-term memory.

Pseudorandom Cue Selection

The section mentions an entirely different approach based on a pseudorandom number generator. Recursive neurons that toggle are available and may constitute various registers and counters, in analogy to classical digital design. It is helpful to have an overview of these in order to understand pseudorandom number generation using toggles.

Pulse Burst Counter

Figure 5.6 sketches a plan for a pulse burst counter.

Synapse S1 serves to convert a first pulse burst into a single pulse that is used to toggle Tog1 from 0 to 1. This produces the least significant bit of a binary count Q_0. The next pulse to Tog1 returns Q_0 back to 0; before this occurs, U1 enables a pulse to Tog2 moving Q_1 from 0 to 1. The result for two stages is a binary count Q_1, $Q_0 = 0, 0; 0, 1; 1, 0; 1, 1$. A two-bit counter element is defined in the figure. These counter elements may be cascaded to produce a count of any number of bits.

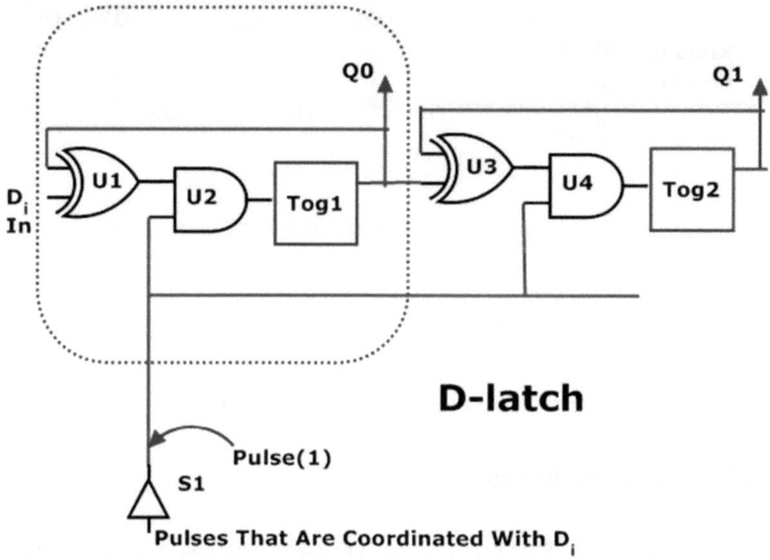

Fig. 5.7 Shift register left-to-right shift register using toggle elements

Shift Register

A shift register forms the basis of a pseudorandom number generator to be described below. This style of generator is well suited to systems that operate without a central clock. However there must be a local sequence of pulses to cause data to shift through the resulting register. Figure 5.7 illustrates a plan.

The D_i are a stream of 0s and 1s that are going to be "clocked" into the register. A sequence of pulse bursts serving as a *clock* are coordinated with the D_i. The clock may be little more than a signal to modify the cues. This signal is converted into an accurate clock pulse via synapse S1.

The shift register works as follows. If D_i is true while Q_0 is false, for instance, then XOR unit 1 applies a true to the AND gate U2. A clock pulse subsequently causes Tog1 to toggle so that Q_0 goes to 1. But if D_i is true while Q_0 is also true, then XOR unit 1 applies a false to the AND gate U2 and no further toggle occurs. Essentially the value of D_i is moved into Q_0.

This also works to shift in a D_i of false. If Q_0 is 0 there is no toggle; if Q_0 is 1 it is toggled to zero. Again the value of D_i is moved into Q_0, giving a shift operation. Similar shift register elements may be cascaded to produce a code of any number of bits. In the field of computer science a binary code is usually shifted into a register in order to store it temporarily. The goal of a pseudorandom number generator is not to store, but to constantly change numbers in a way that appears random.

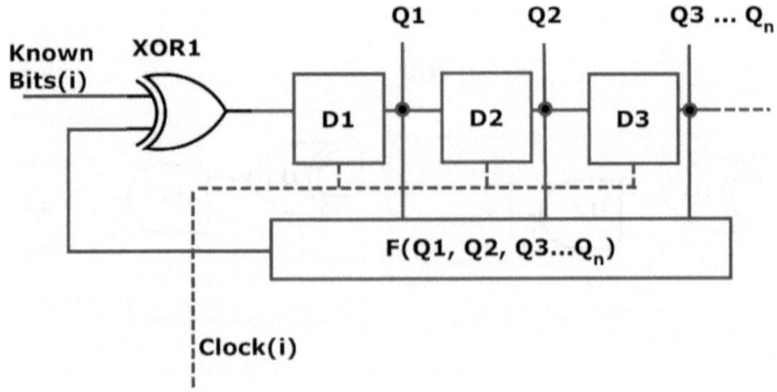

Fig. 5.8 Generalized generator of shift register sequences

Shift Register Sequences

A plan is shown in Fig. 5.8. D1, D2 and D3 are D latches. F is a combinational Boolean function. Bits(i) could be an arbitrary sequence, but here Bits(i) is assumed to be zero for each clock(i) pulse.

A great many sequences can be generated by this particular state machine because there are a great many possible Boolean functions F. For n inputs there are up to 2^{2^n} varieties. This sort of plan is used to test integrated circuit bits, Bits(i). After a given number of clocks, NUM, what results for $(Q_0, Q_1, \ldots Q_n)_{\text{NUM}}$ is known as a signature. If the signature is not exactly correct, the chip is considered to be defective, since it produced one or more wrong bits somewhere. Early electronic calculators were tested this way.

In general the output sequence $(Q_0, Q_1 \ldots Q_n)_i$ is periodic, assuming that the input *Bits*(i) is periodic, meaning that any given sequence will eventually repeat [4]. So it is not truly random. But it has random properties. On average, the 0s and 1s will balance. And for larger numbers there are on average two runs of length R of either 0s or 1s for each run of length $R + 1$ of either 0s or 1s. The codes are not in order and appear random. But technically they are pseudorandom.

A pseudorandom generator can be used to advantage in randomized cue editing. To reduce the above generalized generator to a practical design the function F may be restricted to be "linear":

$$F = c_1Q1 \oplus c_2Q2 \oplus \ldots c_nQ_n. \tag{5.1}$$

Each of the c_i is either 0 or 1. The symbol $\oplus$ means modulo 2 addition, that is, exclusive OR. The resulting period of the pseudorandom sequence will be $\leq 2^n - 1$. For example, if $n = 4$, period will be 15 or less. It cannot be 16 because the code of all zeros, 0000, is excluded. All zeros is not permitted because if the inputs are also all zeros, *Bit*(i) all 0s, no counting occurs. If the sequence has a period of $2^n - 1$ it is defined to have maximal length, which is desirable.

Incidentally, the presence of a maximal length can be predicted. If the following characteristic polynomial cannot be factored, a sequence is maximal length:

$$p(x) = 1 - \sum_{j=1}^{n} c_j x^j. \tag{5.2}$$

For example, $p(x) = 1 - x^2 - x^3$ cannot be factored. So the circuit in Fig. 5.9 should have maximal length.

Note that XOR1 acts like a simple buffer when the inputs are all zero. The shift register needs to be initialized to something other than all zeros. If initialized to 0, 1, 0 the resulting sequence will be as in Table 5.1:

It then repeats. This has a length of 7, which is maximal since $2^3 - 1 = 7$.

Cue selection may be based on a shift register sequence. Figure 5.10 shows a plan.

Shown in part (b) of the figure is an alternative to using a long pseudorandom code; several shorter codes may be used instead. The code is organized as M clusters of code, each with N pseudorandom bits in an attempt to randomize cleverly the selection of attributes. In this way it may be possible to research only those attributes that are unknown.

The shift register is initialized to all ones to begin. Initially none of the cues are masked (blocked by AND gates), so all are permitted to be applied to long-term

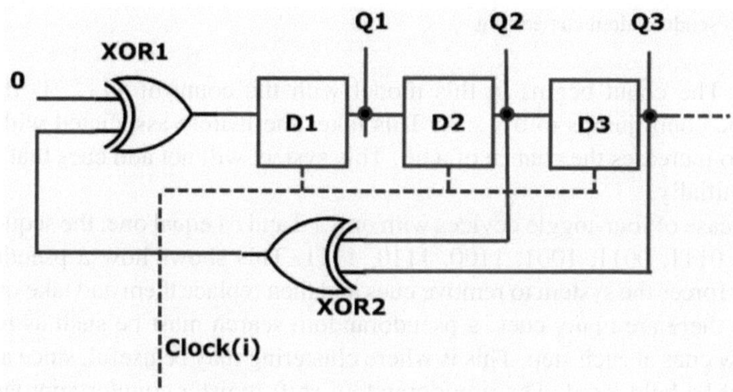

Fig. 5.9 Maximal length pseudorandom number generator example

Table 5.1 Pseudorandom count

010
101
110
111
011
001
100

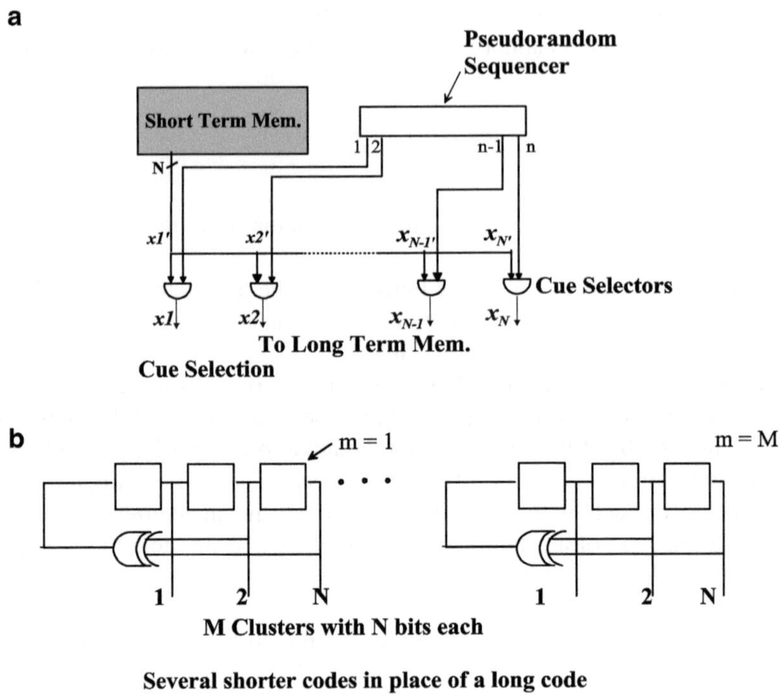

Fig. 5.10 Pseudorandom cue editing

memory. The count begins in this model with the count of 111…1. If there is *NoHit*, the count jumps to 011…1. This takes the feature associated with the "0" out and so increases the chance of a hit. This system will not add cues that were not present initially.

In the case of four-toggle devices with only $c2$ and $c4$ equal one, the sequence will be 1111, 0111, 0011, 1001, 1100, 1110, 1111. This shows how a pseudorandom sequence forces the system to remove cues and then replace them and take others out.

When there are many cues, a pseudorandom search must be such as to change only a few cues at each step. This is where clustering may be useful, since all except one could be held fixed. The pseudorandom shift register counter approach is not elaborated further beyond this point. The counter idea may be useful in understanding, because with two methods, a contrasting is possible.

Conclusions

A cue editor is needed because memory lapses are quite common, and must be resolved fairly quickly, within minutes. Memory search is constantly occurring subliminally, in the background, independent of real time, as evidenced by the fact that recall sometimes takes hours. This supports the proposition of cue storage and circuitry to process cues.

Cues are modeled as being selected from attributes in conscious STM. If they contain conflicts, a match will not be found in long-term memory, and cue editing is desired. A register of simulated qubits is made to accept a given set of conflicting cues and to work on them. Cue editing, however, is not permitted to interfere with the regular processing of consistent cues taken from new images in consciousness.

Simulated qubits change slightly the probabilities for the occurrence of each cue that originally was true with 100% probability. The plan is to remove a few attributes from the list of cues. This increases the chance of associations in long-term memory because fewer cues make association easier. If this does not work, the removed cues are restored and another random set is removed.

Cue editing could also be accomplished with a pseudorandom sequence that removes cues in a random way. Pseudorandom sequencers can be implemented with toggle devices of the sort that recursive neurons produce.

Modernistic concepts have been proposed that could increase the efficiency of cue editing. Particularly useful would be electron tunneling to help synchronize neurons, as mentioned in connection with Walker's work. Also exciting are the possibilities of real qubits, and not just simulated qubits, as mentioned in connection with the proposals of Hameroff and others. The above circuitry could easily be modified to utilize real physical qubits, if in the future they are discovered.

Simulated qubits, although less powerful than the qubits of physics, have abilities that ordinary logic gates do not have, such as holding true and false simultaneously, and giving up their logic values with a probability. This feature is most useful for random cue editing. What is interesting about simulated qubits is that they might exist at the molecular level within a brain, where they would not require a quantum system.

Perhaps more common than conflicting cues would be too few cues, and too many returns. When this happens, it is the task of a *recall referee* to choose the highest priority return for admission into conscious STM, as discussed next.

References

1. Gleitman H (1987) Basic psychology. W W Norton and Co., New York
2. Schacter DL (1996) Searching for memory: the brain, the mind, and the past. Basic Books, New York
3. Schacter DL (2001) Forgotten ideas, neglected pioneers: Richard Semon and the story of memory. Psychology Press, Philadelphia, PA
4. Golomb SH (1967) Shift register sequences. Holden-Day, San Francisco, CA

Self-Study Exercises

(If you desire to keep a record of your efforts, please show a couple of logical steps leading to the answers.)

Cue Editor Logic

1. Concerning Block U4, 5, 6:
   (a) Provide a timing diagram that illustrates the Boolean logical functions involved.
   (b) Comment on pulse synchronization requirements for neural gates. (HINT: Focus only on the AND gate.)
   (c) Explain how the NOT gate works in Blocks U4, 5, 6. (HINT: Use neurotransmitter diagrams as given in a previous chapter.)
2. Provide a schematic to show exactly what is meant by a bus with $K = 5$. (HINT: Show five vertical lines that break out from bus notation with /5.)
3. Are bus conductors most likely to be axons or dendrites? (ANS. Axons are longer and transmit signals faster, so are more probable.)
4. Identify each signal in the proposed cue editor, and its purpose (refer to "Attachment 1" section).
5. Create a timing diagram for Fig. 5.2, the subcircuit for cue editing. (HINT: Assume that all pulses can be appropriately synchronized: 1 ms per gate; 10 ms per simulated qubit; and 10 ms for long-term memory search.)

Probabilistic Simulated Qubits

6. A multivibrator at frequency f_1 produces a pulse width τ. Predict the probability of a true occurring under the following conditions:
   (a) Frequency is at its maximum of f_1 and an ideal sampling pulse whose width is much less than τ is available. (ANS. average probability of a true is 50%.)
   (b) An ideal sampling pulse whose width is much less than τ is available and frequency is reduced by about 10%. (ANS. average probability of a true is about 45%.)
   (c) Repeat the above estimations using a sampling pulse of width of 2τ. (ANS. average probability of a true remains at 50%.)

Pseudorandom Counters

7. Assume that *c2* and *c4* equal one. Verify using a table to trace the outputs of each toggle that the pseudorandom counter of this chapter provides the sequence: 1111, 0111, 0011, 1001, 1100, 1110, 1111.
   (a) Is the sequence maximal?
   (b) Why does not 0000 appear?
   (c) What is involved in designing a code that removes only one attribute at a time? (HINT: Think of a shift register filled with ones in which a zero is shifted across the register.)
8. Design a different 4-bit pseudorandom count sequence using gates as in this chapter.
9. Provide a timing diagram for the major variables for Problem 8.

Attachment 1

Identification of Signals

- Attributes (cues)—Attributes refer to the distributed signals that constitute a mental impression or image. Each attribute for an image is assumed to be held in distributed STM and collectively they represent consciousness. STM is so called because each STM neuron emits pulses that repeat for up to several seconds, thanks to an extended triggering voltage in dendrites and soma.
  The cues that are taken to activate long-term memory are a subset of all possible attributes. The subset is indexed and numbered $1 \leq k \leq K$. The size of the subset K is unknown but it is assumed to be substantial. The value K represents the number of communication channels going to associative (long term) memory. It is also the number of simulated qubits assumed to aid memory search in the event of inconsistencies in the cues.
- *CueMods*(1)—This signal goes true to permit modifications of the cues and a new search, and also clearing of *R1* when a match is returned. *CueMods*(*1*) can go true only if the register flag is set and if there are no new searches that need to be done.
- *New*(1)—This is a signal that indicates when a new image is being applied to STM. *New*(*1*) is a regular (shorter) burst of pulses, perhaps ten pulses. It will enable a search of long-term memory and will also prevent any simultaneous searches stemming from the simulated qubits that are dealing with past inconsistent cues.

- *New*(i)—The index "*i*" is assumed to begin at 1 and to increase indefinitely as the number of images in STM increases. *New*(*1*) is thought of as the first image or impression of the day.
- R1—This refers to a register of *K* simulated qubits plus an extra such qubit that serves as a flag to signal that *R1* is in use.
- Clear—This is a signal that will effectively clear the register so that it is prepared to accept another batch of conflicting cues.
- Flag—This signal indicates that a set of cues has been loaded into the register of simulated qubits. These bad cues are about to be modified.
- *NoHit*(1)—This is a signal from associative long-term memory that indicates that a match was not found. *NoHit*(*1*) causes *R1* to be loaded with the problem cues, but only if the *NoHit*(*1*) occurs shortly after a search attempt via *New*(*1*). Thus, if *NoHit*(*1*) goes true as a result of an attempt from *R1* to find a match, no additional loading takes place.
- *Hit*(1)—This signal goes true when there is a match, which will be the case most often. When *Hit*(*1*) $=$ 1 and *R1* contains data *R1* is cleared to be ready for the next set of problem cues. Note that *Hit*(*1*) and *NoHit*(*1*) are brief neural bursts that normally are both at rest until after a search occurs.
- *Load*(1)—When this signal goes high (true) the simulated qubits are enabled to load the applied cues.
- *Modify*(1)—When this signal goes true the probabilities of having a given cue are reduced slightly to enhance the chance of finding a match in long-term memory.
- U4, 5, 6—This is a block of logic, defined above, that serves to disable a search based from *R1* when there is a need for immediate service of a new set of attributes, that is, when *New*(*2*) goes true.
- U1, 2, 3, 7, 8, 9, 10—These refer to units of neural logic including ordinary AND and OR gates.

Chapter 6
Plans for a Recall Referee

Introduction

It is amazing to think that sophisticated digital processing might be possible using neurons. Yet it is suggested below that digital processing is indeed possible, thus avoiding as much as possible a need for inaccurate and error-prone analog processing. Brain logic, although a good thing, is chiefly asynchronous and therefore considerably more complex than computers with central clocks. Central to the digital concept are simulated qubits operating as controlled toggle switches.

Quite often, it seems, cues are underdetermined, in which case several possible images will come forth in succession from long-term memory. Returns, assuming more than one, do not go directly into conscious short-term memory, but are directed into registers of simulated qubits. Registers, like words, are a logical concept; the exact physical form of the circuitry remains unimportant.

Each return comes from a word of memory, and moves into a given register of simulated qubits. In parallel, the registers undergo a calculation of priority. Suggestions for establishing priorities have been proposed in connection with artificial intelligence [1, 2] and these have been well discussed [3].

Attributes are assumed to be identified by their position in distributed memory and this position implies a certain weighting factor for priority. This book does not specify the weightings of attributes for a priority calculation, but surely the computation will contain survival factors, emotional factors, and also sensory intensity (bright light, loud sound, etc.).

The author assumes a binary weight with N_1 bits (N_1 is a variable, not large; the following example uses $N_1 = 3$). It is reasonable to assume a higher weight for attributes that relate to physical danger, for example binary 100. The next highest could be for emotions, perhaps 010. A lower weight could be for intensities relating to what was sensed when the memory was created, say 001. To establish a priority for each image in the simplest possible way, these weights are added (digitally) to determine priority.

J.R. Burger, *Brain Theory From A Circuits And Systems Perspective*, Springer Series in Cognitive and Neural Systems 6, DOI 10.1007/978-1-4614-6412-9_6,
© Springer Science+Business Media New York 2013

Note that binary weights are restricted to insure that intended priorities are computed. For instance, danger creates a higher priority, 100 plus something. The other weights sum to only 011, and are chosen so that they will never overrule danger.

Addition is accomplished with the help of code stored in a special portion of subconscious long-term memory. The stored code directs a meaningful sequence of controlled toggles. Signals from given *source* qubits instigate toggles in given *target* qubits, ideally disjoint from the source qubits. Simulated qubits can communicate with each other as appropriate since there are interneurons that serve as a bus.

A system like this is logically reversible, with implications for energy efficiency. The system is capable of adding numerical weights, plus a wide variety of other exotic arithmetic beyond the scope of this chapter. Real qubits, by the way, would not necessarily need a conductive bus, but require external forces, usually electromagnetic, to manipulate states, effectively permitting qubits to communicate without a bus. As with simulated qubits, signals from given *source* qubits are thus able to instigate toggles in (different) *target* qubits.

Priorities for multiple returns are going to be calculated in parallel to make recall as fast as possible. Parallel computing is one of the advantages of controlled toggles.

The brain system is basically associative [4] and is driven by returns. When returns occur, an *Enable Calculations* signal goes true; priorities are then computed and compared. Soon after, the highest priority image is gated into consciousness.

Overview of a Recall Referee

A recall referee may be conceived as a dedicated block that does only one thing, compute priorities. Figure 6.1 is an overview of a possible system. Returns flow from long-term memory, and are directed automatically into registers of simulated qubits. A given return is assumed to have up to K attributes, $1 \leq k \leq K$, where K is an unspecified variable. Each and every image in long-term memory is made up of differing standardized attributes, and of course a given image would never use all available attributes, so those attributes processed are well below K in number.

A signal labeled *Hit* from long-term memory indicates that an image is being returned; it will be directed into an available *Attribute Register*. It is assumed that there are P registers (P is not specified here).

The important attributes are going to be encoded digitally and entered into a *Toggle Register* as suggested in the figure. An upper limit of qubits per register will be the number of attributes being processed times the number of bits per attribute, $K \times N_1$. However, not all attributes need to be encoded, so in fact the number is well below $K \times N_1$. Additional scratchpad qubits are needed for calculations, so a grand total of L simulated qubits are involved (L is not specified here).

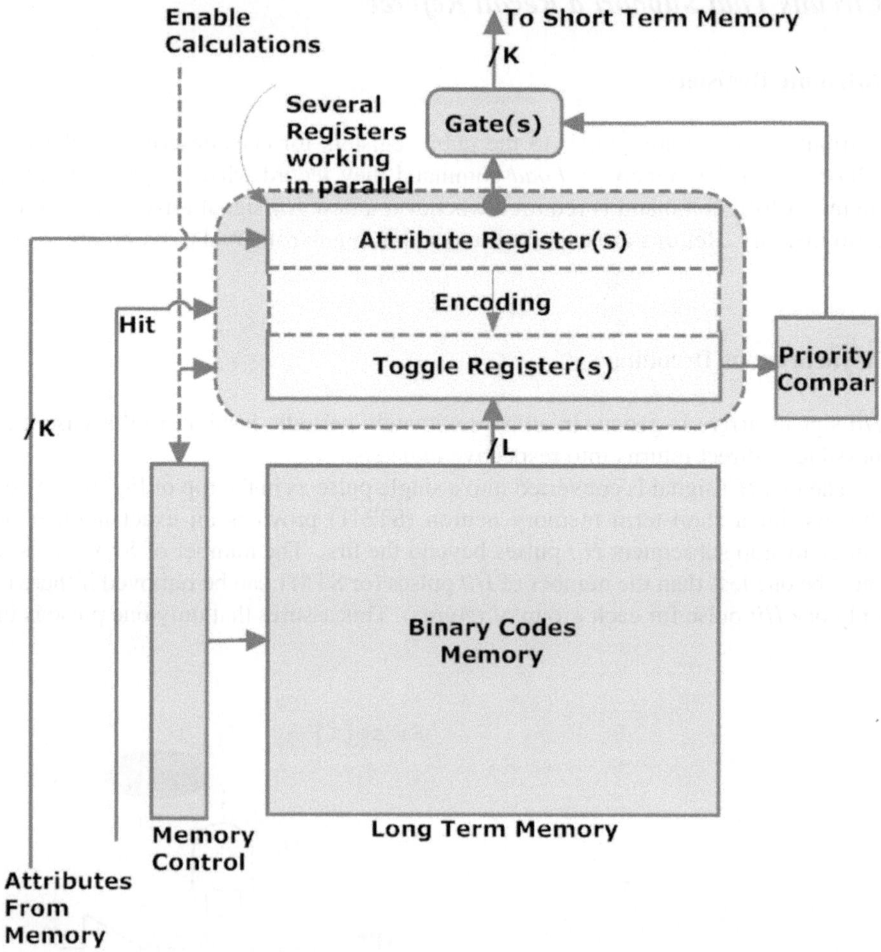

Fig. 6.1 Recall referee overview

A signal labeled *Enable Calculations* will begin the addition. Addition using controlled toggles belongs to the field of reversible computing and is explained in a later chapter.

Controlled toggles are carried out under codes stored in a block of long-term memory. These codes apply to each toggle register in parallel in order to compute a priority for each register in parallel.

To begin, the *Enable Calculations* signal which is a single pulse will ripple through code memory. When it reaches a word filled with code, it applies the code to the toggle registers. In this way each code is applied sequentially to the toggle registers where the calculations are being accomplished in parallel, piece by piece. Finally the priorities, *P* in number, are ready, and are compared in a *Priority Comparator*. The highest priority image is subsequently gated into consciousness.

Circuits That Support a Recall Referee

Attribute Registers

Attribute registers are similar to the qubit registers for cues described under cue editor. When they receive a *Load* command they record what is applied to their inputs. A load command is required whenever a new *Hit* signal arrives from long-term memory. Returns are expected to occur at regular intervals as a group.

Asynchronous Decoding

Hit signals arrive in groups in an approximately periodic fashion. In this case it is possible to direct returns into respective registers.

The first *Hit* signal is converted into a single pulse as in the top of Fig. 6.2. After the first hit a short-term memory neuron (STM1) provides an exact number of pulses to stop subsequent *Hit* pulses beyond the first. The number of STM1 pulses must be one less than the number of *Hit* pulses (or STM1 can be removed if there is only one *Hit* pulse for each group of returns). This assures that only one pulse is in

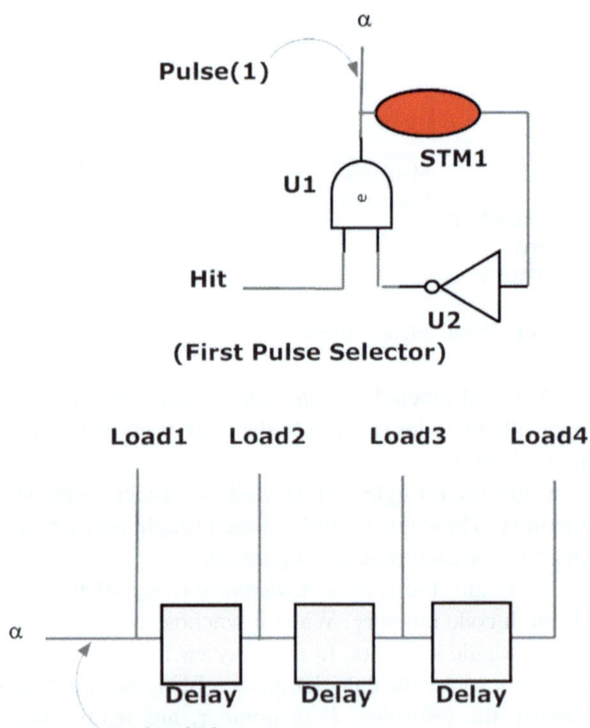

Fig. 6.2 Generation of load commands

Fig. 6.3 Attribute encoding

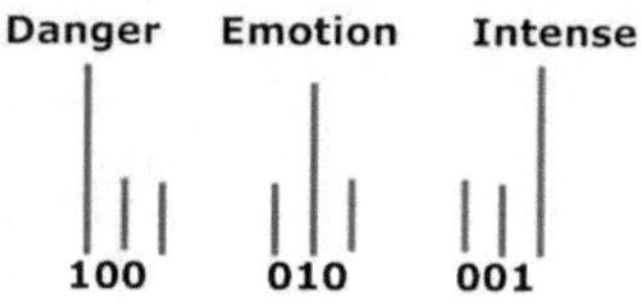

the delay chain in Fig. 6.2 for a given group of returns. The single pulse ripples through the chain, sequentially activating the individual load signals.

Encoder

Once an image is in a register, binary weighting is expected. For instance, assuming precision $N_1 = 3$, a "danger" attribute could translate to 100 as portrayed in Fig. 6.3. Other important attributes are translated similarly.

These codes could be intrinsic to attributes in memory. Alternately, they might be automatically generated by virtue of the location of an attribute, since location is assumed to define what an attribute means. Not all attributes need weights. Those whose weights are low are set to zero.

Toggle Registers for Computing Priority

Toggle registers are controlled by codes from long-term memory. The codes contain fields labeled *fm* and *to*. These will be applied to toggle circuits with connections as in Fig. 6.4. The circuits are arranged so that if all of the toggle elements are true in a given *fm*, the conductor bus will be at rest. But if any one of the toggle elements is false, the bus is made true. The bus, in this implementation, is a single interneuron. This method is inspired by a hardware design (Burger patent number 7,242,599, July 10, 2007).

When the bus is at rest, the toggles in a given *to* field are flipped, true to false, or false to true. For example, say the output of **T2** is true. If a *fm* signal is directed to **T2**, then a false is held to the bus by gate U4. So the bus is at rest. If a *to* signal is directed to **T1**, the bus false is inverted to true at the input of gate U1, the output of which directs **T1** to toggle.

As another example, say a source *fm* addresses **T1** and **T2**; the complement of the output **T1** and the complement of the output **T2** are held to the bus. If both the output **T1** and the output **T2** are true, a false appears on the bus. In other words, if the AND of the outputs **T1** and **T2** is true, a false is held to the bus; this means the bus is at rest. So those toggles with a to-signal will flip states.

But if one of or both **T1** and **T2** contain a false at the output, pulses will appear on the bus. If there are pulses on the bus, the destination toggles identified by the *to* will not toggle. Multiple *fm* signals control multiple *to* toggles.

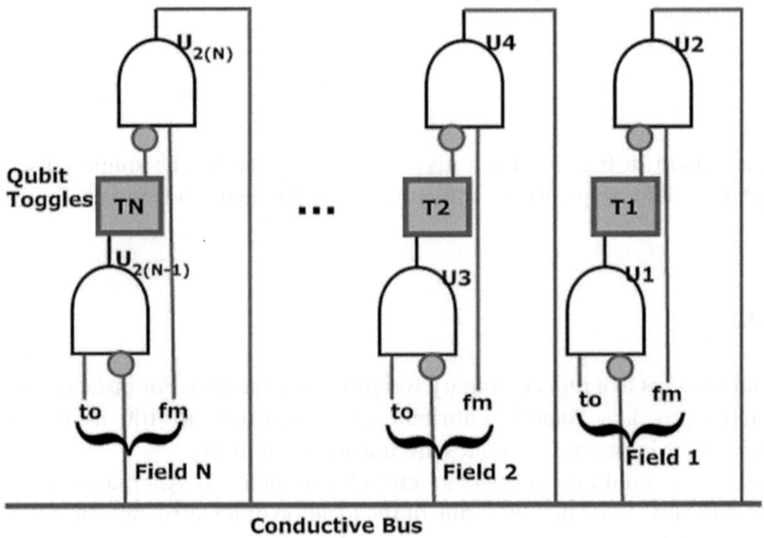

Fig. 6.4 Structure of toggle registers for computations

This is an example of *wired logic*, similar to that of hardware buses using open drain or open collector transistors.

A programmed toggle register may compute a wide variety of mathematical functions. Here it will be arranged to sum the weights associated with attribute encoding, the details of which are left to another chapter. An advantage of this method is that several images in several register are going to compute priorities in parallel.

Read Only Memory System

Running the codes that control the toggle registers is straightforward. They reside in a special section of long-term memory, if not a special molecular memory packet. Assuming long-term memory, the code at the top of a code stack is applied first, then the next, and so on until the end of the program. A counter and decoder are not required if the codes are fed periodically to the toggle registers at a proper rate. Figure 6.5 shows how this might be accomplished.

Code 1, carried by a neural bus of L paths, is applied first through U1. Then, after a delay, code 1 is disconnected and code 2 is applied through U2. This ripple process is repeated until the end, code Z. The various OR gates serve to transmit the

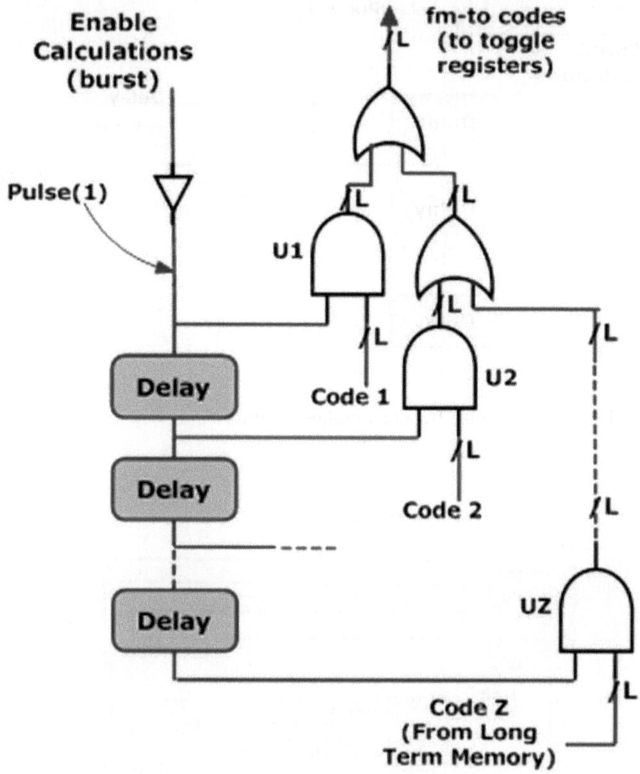

Fig. 6.5 Applying *fm–to* codes one at a time

codes upward. A chain of two-input OR gates is shown. Alternatively, neurons may create a single OR gate with many inputs.

The codes come from a special section of long-term memory, probably inborn. Their use, of course, is subconscious.

Toggling of simulated qubits for priority calculations does not require a random sampling. A sampling pulse may be generated from the same delay system that reads out the *fm–to* codes. This part of the circuit is as in Fig. 6.6.

Priority Selection

This refers to choosing that returned image with the highest priority for entry into consciousness. Figure 6.7 shows how this might be done.

The priorities are assumed to be computed (in binary form) and applied to a magnitude comparator. The outputs will all be false except one, the one with the

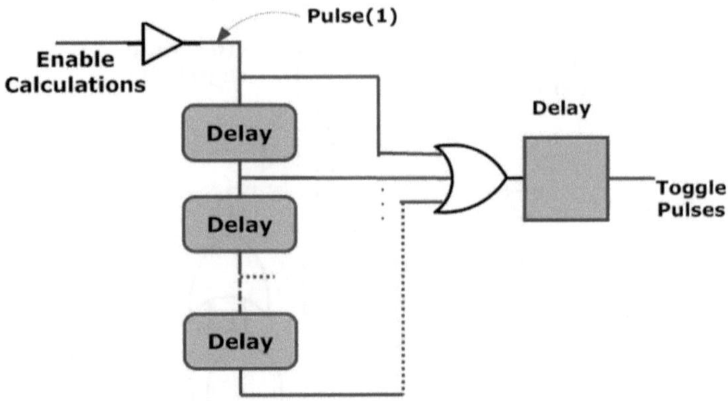

Fig. 6.6 Producing a toggle pulse for the simulated qubits

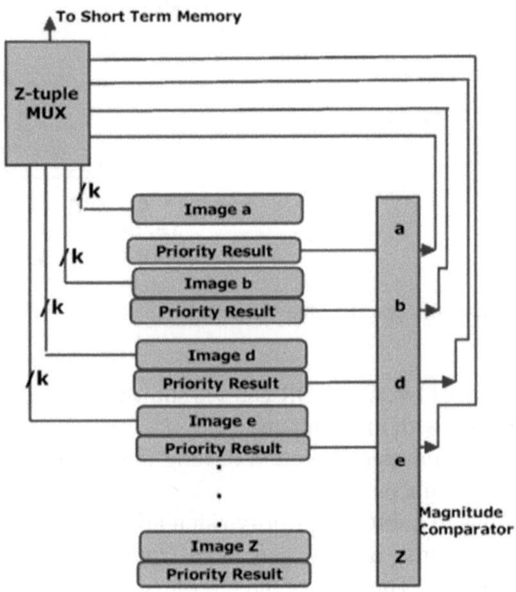

Fig. 6.7 Gating higher priority image into consciousness

highest priority. The output of the comparator is assumed to be decoded so that only one output is true. This assures that only one corresponding image is presented to short-term memory.

The multiplexer (MUX) is a high-level switch that switches *k*-tuples of image attributes. A basic MUX circuit is shown in Fig. 6.8.

Fig. 6.8 Image multiplexer

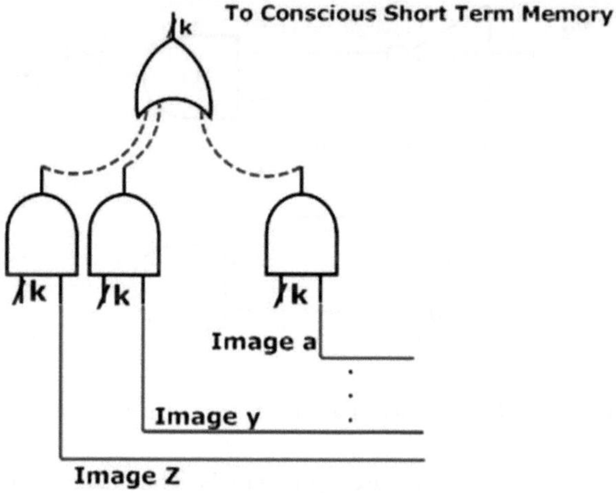

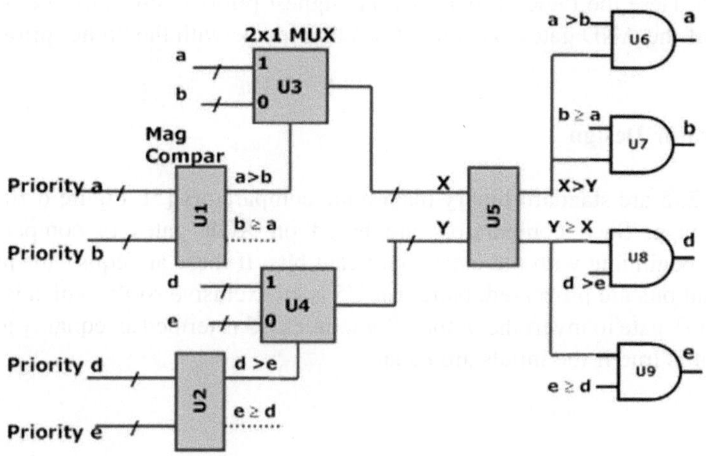

Fig. 6.9 Comparator for four priorities

Priorities' Magnitude Comparator

A magnitude comparator may be implemented using *fm–to* codes applied to a toggle register, as suggested in a later chapter, but here is a design using random logic. Priorities may be compared two at a time as in Fig. 6.9.

This example involves four calculated priorities. Pairs of priority values are compared in U1 and U2, which are comparison blocks. The true or false outputs of these blocks are $a > b$ or $a \leq b$ and $d > e$ or $d \leq e$. The symbols $a > b$ or $a \leq b$

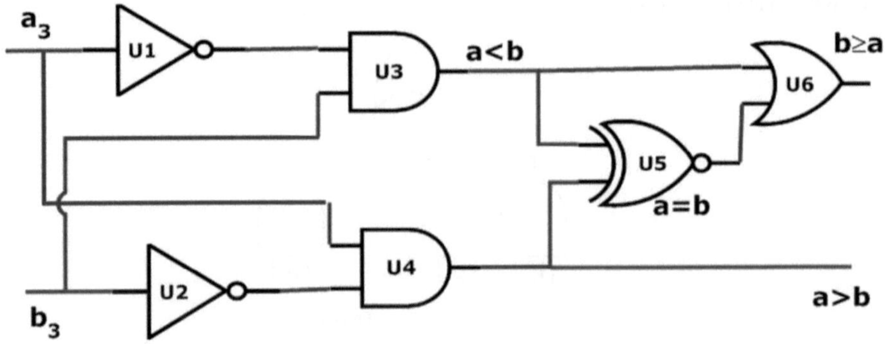

Fig. 6.10 Typical stage of a magnitude comparator for bits a_3, b_3

and $d > e$ or $d \leq e$ are signal labels. They denote a physical label-to-label connection even though a path is not shown (portrayed this way for clarity).

The outputs $a > b$ and $d > e$ control a 2×1 MUX, similar to the MUX above. This MUX is connected to send the highest priority values to a third comparison block U5. Here the logic is such that the highest priority forces to true one of the outputs of the AND gates. This identifies that image with the higher priority.

Comparator Design

Units 1, 2, 5 are standard binary magnitude comparators [5]. Figure 6.10 shows a typical stage. Such comparators are based on XOR gates to compare binary numbers beginning with the most significant bits. If these are equal, the next most significant bits are processed. Note that U5 is an exclusive NOR, which is an XOR with a NOT gate to invert the output. Sometimes U5 is termed an equality gate since its output is true if the inputs are equal.

Timing Estimations

If there is only one associative return, there is no need for a priority determination. In this case the image is gated into short-term memory immediately (provisions for this are not shown). Going through associative memory likely requires only a few tens of milliseconds.

Gamma brain waves display synchronized oscillations in the range of 30–90 Hz and are a possible correlate of consciousness. A rate of images moving through conscious short-term memory might be taken to be 40 Hz. This means a refreshed image about every 25 ms, although no doubt this can vary considerably. A cycle time of a few tens of milliseconds, say 25 ms, is comparable with the basic rate in the above system.

With multiple associative returns and a need for priority calculations and selection, cycling is markedly slower. A small amount, perhaps 30 or 40 ms, is taken for editing and memory access. Using 2 ms per neural gate, and 3 bits of addition for priority, there is roughly 300 ms results for priority selections. Consciousness would, in a worst case, come in flashes at a rate of one every 1/3 s approximately. The stream of consciousness is not smooth under such adverse conditions, but nonetheless is adequate for most real-world responses.

Enable Calculation Signal

This signal occurs whenever a group of associative returns are standing by waiting for processing. This implies that returns can be identified as a group so that when the entire group is available, the processing begins. The author visualizes a digital filter that detects the end of a group of returns. This could signal the beginning of a priority calculation (see Fig. 6.11 for the case of three returns maximum in a group). This approach is easily expanded.

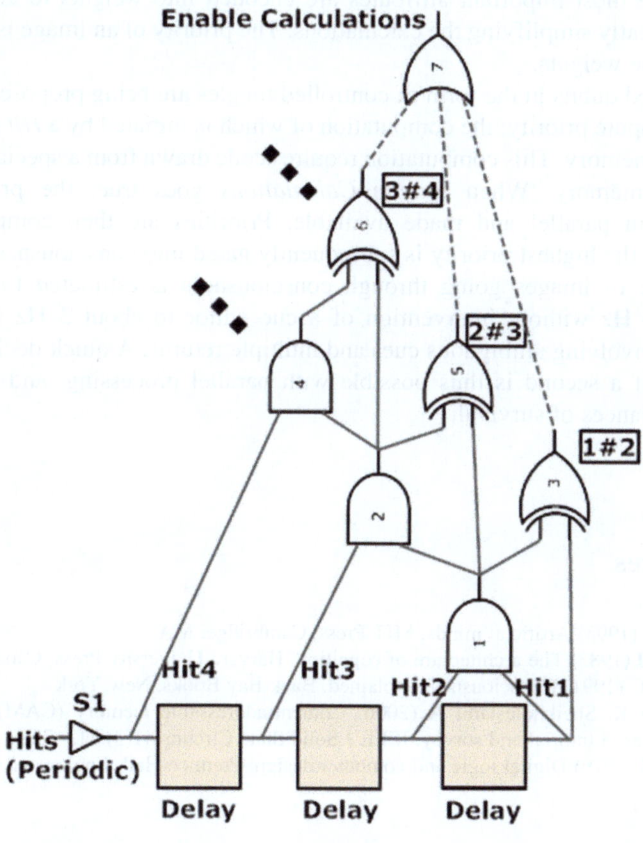

Fig. 6.11 Detecting the end of a group of returns for the case of three returns maximum

This approach assumes that a group of hits come periodically. The weak synapse S1 is to provide a single pulse for accuracy. Delay is fixed to match the times between multiple returns. The symbols 1#2 means one return but not two; 2#3 means two returns in a group, but not three; 3#4 means three returns in a group, but not four. When a group ends, an enable calculations signal is emitted.

For three hits in a sequence, Hit3, Hit 2, and Hit 1 are true; Hit 4 is false; output of U1, U2, and U3 is true; output of U4 is false. So U6 has a true output that is transferred through U7 to provide an enable calculations signal.

Conclusions

Over countless millennia, many billions of neurons and many trillions of synapses have evolved. Electrical engineers and others have learned how to deal with circuit complexity as expected in a brain. One approach is to model calculations as arrays of controlled toggles.

Only the most important attributes are encoded into weights to contribute to priority, greatly simplifying the calculations. The priority of an image is simply the sum of these weights.

Simulated qubits in the form of controlled toggles are being proposed as a good way to compute priority, the computation of which is initiated by a *Hit* signal from long-term memory. This computation requires code drawn from a special section of long-term memory. When *Enable Calculations* goes true, the priorities are computed in parallel and made available. Priorities are then compared. That image with the highest priority is subsequently gated into consciousness.

The rate of images going through consciousness is estimated to vary from roughly 40 Hz without intervention of a cue editor to about 3 Hz for difficult situations involving ambiguous cues and multiple returns. A quick decision within one-third of a second is thus possible with parallel processing, and enhances a person's chances of survival.

References

1. Franklin S (1995) Artificial minds. MIT Press, Cambridge, MA
2. Anderson J (1983) The architecture of cognition. Harvard University Press, Cambridge, MA
3. Dennett DC (1991) Consciousness explained. Back Bay Books, New York
4. Pagiamtzis K, Sheikholeslami A (2006) Content-addressable memory (CAM) circuits and architectures: a tutorial and survey. IEEE J Solid State Circuits 41(3):712–727
5. Mano MM (1979) Digital logic and computer design. Prentice-Hall, Englewood Cliffs, NJ

Self-Study Exercises

(If you desire to keep a record of your efforts, please show a couple of logical steps leading to the answers.)

Simulated Qubit Registers

1. Provide a timing diagram for the Fig. 6.2 First Pulse Selector given in this chapter.
2. Provide a timing diagram for the Fig. 6.2 Load 1, 2, 3, 4 signals as in this chapter, given a single pulse input, and assuming a delay of 10 ms.

Encoding

3. Using the encoding of Fig. 6.3, calculate priorities for the following combinations:

   (a) Danger, Intense (ANS. 101)
   (b) Danger, Emotion (ANS. 110)
   (c) Danger, Emotion, Intense (ANS. 111)
   (d) Emotion, Intense (ANS. 011)
   (e) Which combination has the highest priority? (ANS. c)

Toggle Register Codes

4. For the circuit of Fig. 6.4 explain what the following codes accomplish:

   (a) 01...0010
   (b) 10...0101
   (c) 10...1001

5. For Fig. 6.4, is there a problem with the code 00...1100? Note that Field 2 is both 1s. Explain! (HINT: The output of toggle 2 is used to control the input of toggle 2. This sort of operation is logically irreversible, since toggle 2 resets to zero; it can never return to one without better code. Also, if the *fm* signal is not removed in a timely way there will be extra energy loss as the bus propagates a true level.)

Code Transfer

6. Refer to Fig. 6.5. Redesign this circuit so that it uses a single many-input OR gate instead of a chain of two-input OR gates.

Priority Selection

7. Refer to Fig. 6.8. Redesign this circuit so that it uses only two-input OR gates instead of one large many-input OR gate.
8. Apply the following numbers to the Fig. 6.9 four-priority comparator and determine which gates pass a signal to the output. Mention as required U1, 2, 3, 4, 5, 6, 7, 8, 9.
   (a) a, b, d, e are 011, 100, 010, 101
   (b) Repeat for 101, 010, 111, 111
9. Refer to Fig. 6.10. Use a truth table showing all internal variables, that is, each of the outputs of units U1–U6 as the inputs a_3 b_3 count up in binary; trace the logic possibilities and to prove that this comparator works to compare two bits as advertised.

Enable Calculation Signal

10. Explain with a timing diagram how the logic works in Fig. 6.11 to generate an *Enable Calculations* signal.

Chapter 7
Arithmetic Using Simulated Qubits

Introduction

Arithmetic is generally accomplished by continually swapping the contents of local registers and random access memory, running all numbers through a central processor to build a result. Unfortunately this is inefficient, and impractical in a neural environment. The register-stationary manipulations of controlled toggles are far more efficient, plus they are biologically convenient. Registers of simulated qubits operating as controlled toggles also happen to be logically reversible, with implications for lower energy dissipation. Controlled toggles directed by codes taken from long-term memory are well suited to massively parallel brain computations.

Simulated qubits imitate, to some extent, the logical nature of real qubits. In particular, with the aid of interneurons, they may support a system of unconditional toggles (UT), single controlled toggles (SCT), and multiple controlled toggles (MCT).

Physics parlance has used the term "NOT" in place of the term "toggle." To them, for example the UT is called UN, or unconditional NOT. Names are important; so for the record, NOT refers to an inverter whose output depends on a nonlinear way on a steady input such that the output maps to the input. A NOT gate has no memory whatsoever.

In contrast, a toggle device is a memory device; it remembers its most recent value. A toggle has no input although it is innately capable of being triggered by a brief pulse to change states. Once established, output is held indefinitely, or at least until power is removed. The differences between a NOT and a toggle are fundamental.

Each simulated qubit may be configured to be in a given logic state with 100 % probability. A toggle means a switch to the opposite logic state with 100 % probability. Thus, true goes to false, and vice versa.

Image priority follows by adding the weighting factors that are associated with important attributes in an image, such as indicators of danger. Calculations are

J.R. Burger, *Brain Theory From A Circuits And Systems Perspective*, Springer Series in Cognitive and Neural Systems 6, DOI 10.1007/978-1-4614-6412-9_7,
© Springer Science+Business Media New York 2013

contained wholly within the register of simulated qubits that holds a given image, and this is where priority will be stored. Addition is well understood, but when it is constrained to be accomplished locally, without external memory and data buses, details are important.

Special terms are helpful in order to describe how controlled toggles can be programmed. Qubits that are being toggled are termed "*to*" qubits. The qubits that are controlling are termed "*fm*" qubits. Logic is such that all controlling *fm* qubits must be true in order to toggle all *to* qubits. Usually the controlling qubits are disjoint from those qubits being toggled, since this preserves logical reversibility.

Controlled toggles are sufficient to accomplish a wide variety of digital arithmetic without a power-dissipating data bus. This method is powerful; geniuses and savants might very well rely on controlled-toggle operations performed subconsciously on a massively parallel scale. However the main purpose of this chapter is not arithmetic in general, but to show how weighting factors may be summed digitally to compute priority. Each returned image has its own register where priority is calculated, all in parallel.

Symbols for reversible logic are essential and will now be introduced.

Introduction to Controlled Toggling

A register of controlled toggles can be mapped as in Fig. 7.1.

This word will be given an ability to test a *selected* set of its own q-qubits, defined to be the *fm* qubits; if the tested qubits are all true, then one or more objective bits within itself, the *to* qubits will be complemented or flipped.

At this point it is helpful to define a "wiring" diagram. Wiring diagrams were introduced by reversible computing researchers such as Charles H. Bennett [1]. A wiring diagram does not involve any wires; rather, a wiring diagram serves to specify logically reversible programming. One may imagine that each line in a wiring diagram is related to a simulated qubit position in each of the parallel register words as suggested in Fig. 7.2. There may be any number of words, up to $L = 2^N$, each with different bit patterns. Each word is going to be modified

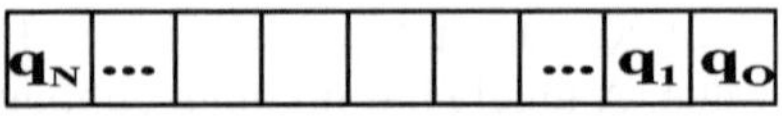

Fig. 7.1 Toggle register

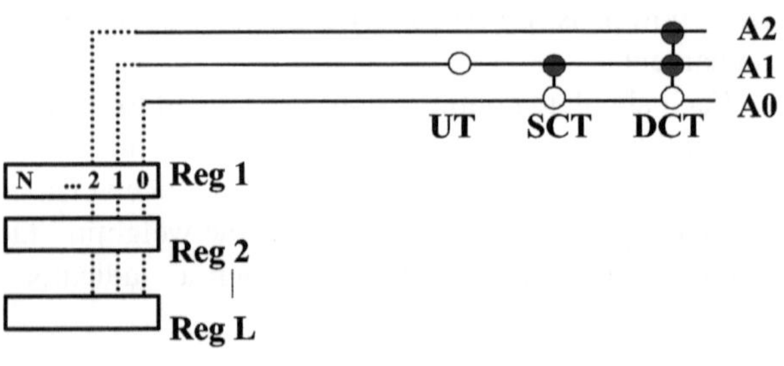

Fig. 7.2 A wiring diagram with example symbols and its relationship to toggle positions

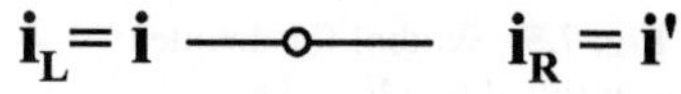

Fig. 7.3 Symbol for unconditional toggle or UT

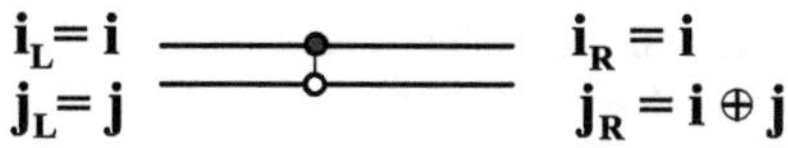

Fig. 7.4 Symbol for single controlled toggle, or SCT

according to the instructions in a wiring diagram, so in a sense the wiring diagram is the program.

One usually visualizes registers to be horizontal; the flow of register operations is usually imagined as going from left to right. A wiring diagram maintains this concept.

Shown in the figure are the symbols for UT, SCT, and double controlled toggle (DCT); these and other symbols, such as MCT, have been referred to as "gates." The term "gate" is misleading, as mentioned above, given that they are actually toggles of various types. Figure 7.3 illustrates an unconditional toggle:

This applies to a simulated qubit which is in a true or a false state i such that a logic value $i_{\mathrm{L}} = i$ on the left appears on the right as $i_{\mathrm{R}} = i'$ meaning the NOT of i. This means that upon triggering, or application of a UT, a toggle occurs that is not controlled by other signals, that is, an unconditional toggle. The wire is really a sort of timeline from *left to right*, denoting before a trigger (on the left), and after a trigger (on the right).

CMOS, for instance, might work from left to right but does not work backwards. What differs from conventional logic is that a timeline of a wiring diagram may go from *right to left* so that time, in a logical sense, can be reversed if need be. If i' is applied on the right, then i appears on the left; this gate is *logically reversible.* Reversibility has implications for data recovery and for reducing wasted energy (calories). The *bubble* in a wiring diagram is the symbol for the unconditional toggle.

Figure 7.4 illustrates a single controlled NOT, or SCN:

There are two toggle switches involved, each in true or false states i and j. In this case the input $i_{\mathrm{L}} = i$ on the left is simply carried through with no change so that on the right $i_{\mathrm{R}} = i$. The little black *dot* is a standard symbol indicating a connection, meaning that toggle i controls toggle j. The $j_{\mathrm{L}} = j$ on the left is complemented only if i is true. Another way of expressing what is accomplished by an SCN is

$$j_{\mathrm{R}} = i \oplus j.$$

The symbol $\oplus$ means XOR of i with j, that is, XOR(i, j). If i is false, then $j_{\mathrm{R}} = j$, and nothing happens. If i is true, then $j_{\mathrm{R}} =$ NOT(j), that is, j'. Inputs can be applied to the right as readily as to the left, so the SCN is logically reversible. Note that the input i is carried through unchanged and assures that information is not lost (as it would be in irreversible logic).

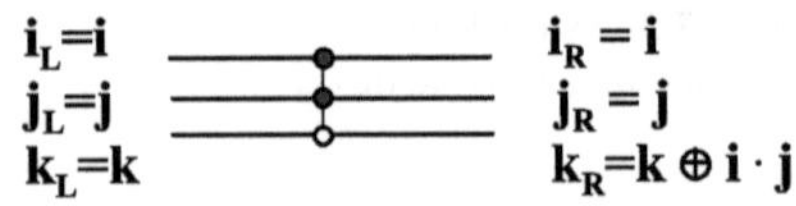

Fig. 7.5 Symbol for double controlled toggle or DCT

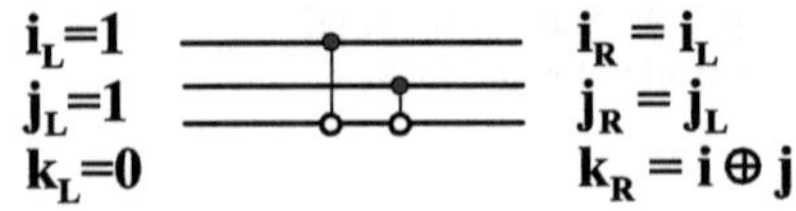

Fig. 7.6 Reversible modulo-two addition of $i + j$

The idea of controlled toggles can be extended to any number of controlling lines, creating what may be termed a DCT. The symbol for DCT appears in Fig. 7.5.

The double controlled NOT computes the AND of i and j; if this combination is true, then $k_{\mathrm{R}} = \mathrm{NOT}(k_{\mathrm{L}})$. The i and j are transmitted without change so that $i_{\mathrm{R}} = i_{\mathrm{L}}$ and $j_{\mathrm{R}} = j_{\mathrm{L}}$, to ensure reversibility. The DCT is also known as the Toffoli gate [2]. It is a special case of a *MCT* which depends on the AND of multiple controls. MCT are useful for detecting when a group of inputs all equal one.

Simple Applications

As a simple application, imagine many registers as in the above figure, one of which contains a hidden code. As a simple example, $q_2 = 1, q_1 = 1$ could be hidden. The idea is, this code points to secret information. Registers can be analyzed concurrently using a DCN. First initialize $q_0 \rightarrow 0$ in each register. Then apply the DCN to qubits q_1, q_2 where the code is hidden. That word that contains the code $q_2 = 1$, $q_1 = 1$, $q_0 = 0$ on the left of the wiring diagram will convert to $q_2 = 1$, $q_1 = 1$, $q_0 = 1$ on the right of the wiring diagram. In other words, the flag in the form of $q_0 = 1$ is raised to identify the hidden code. Thus the secret information the hidden code points to has been located and may be observed.

On a much larger scale the application of MCN suggests a solution to a *needle in the haystack* problem.

As an example of logically reversible adding, consider modulo-two addition of 1 + 1 as in Fig. 7.6. Modulo-two addition means that the sum must show either 0 or 1, but not 2. Therefore 1 + 1 = 0, with the implication that there is a carry of 1 and a sum equal to 10, or binary 2.

This circuit computes the modulo-two sum, not the carry. Assume three toggle devices in states i, j, k. Assume that k on the left is initialized to $k_{\mathrm{L}} = 0$. The modulo-two sum is computed in k_{R} on the right. For example, if a word had on the left $i_{\mathrm{L}} = 1$, $j_{\mathrm{L}} = 1$, and $k_{\mathrm{L}} = 0$, then on the right is $k_{\mathrm{R}} = i \oplus j = 0$, the correct answer. The modulo-two sum is an exclusive OR. Using wiring diagrams with more complexity, reversible addition may be extended any number of bits.

It is important to note that data in the above registers does not actually move out of the register, as would be necessary in microprocessors. This processing is

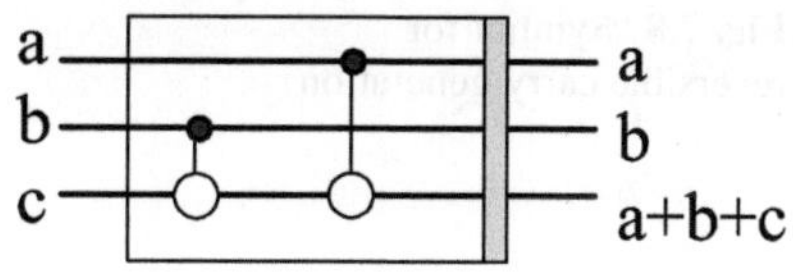

Fig. 7.7 Symbol for reversible modulo-sum generation

register-stationary. Typically in man-made hardware, data must be moved from disk memory to local random access memory and then to various registers in a microprocessor; then it must be moved back to local memory and then back to disk memory.

Random access reading and writing is the mainstay of desktop computers. But this dissipates extra heat and uses extra time, and a bottleneck occurs on the data bus that drastically slows down a conventional computer. The biological brain can avoid such inefficiencies.

Reversible Addition of Positive Integers

A reversible adder will now be analyzed as having two components: a block that computes the sum, and a block that computes the carry.

Reversible Sum

A modulo-two sum of three bits may be computed logically as in Fig. 7.7. The block receives the binary inputs **a**, **b**, **c** and produces the binary outputs ***a***, ***b***, ***a*** + ***b*** + ***c***, where + in this context means modulo-two addition, symbolized as follows.

Note that c can be thought of as a carry in, and will be toggled if $a\, a \oplus b = 1$. To preserve reversibility, a and b are unchanged. For example, let $a, b, c = 0, 1, 1$; then, out of the right side comes 0, 1, 0. The a + b + c equals zero, the modulo-two sum of the three bits 0 + 1 + 1. If $a, b, c = 1, 1, 1$, then, out of the right side comes 1, 1, 1. The a + b + c equals one, the modulo-two sum of 1 + 1 + 1.

It is necessary to specify the direction of the flow. The bar in the block symbol indicates the direction of the flow, from left to right, so operations execute in order from left to right. This calculation is logically reversible. Entering 0, 1, 0 on the right, for example, and executing gates in order from right to left result in 0, 1, 1 on the left. No information is lost since reversible toggles are used.

Reversible Carry

A carry c_1 may be computed as $\boldsymbol{c_{in0}\, a + c_{in0}\, b + a\, b}$, where + in this context means OR. This is known in digital design as the *two-out-of-three* function. The three

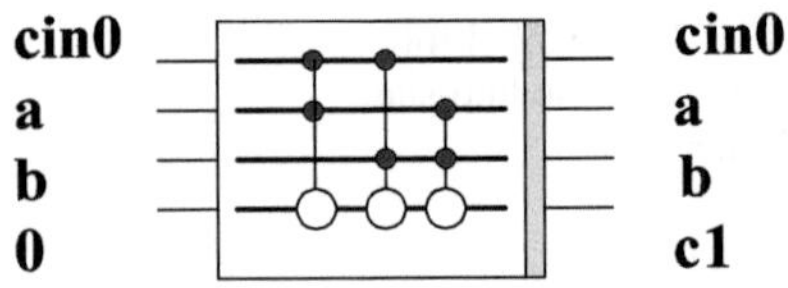

Fig. 7.8 Symbol for reversible carry generation

Table 7.1 Binary addition example

(1 1 1)
1 1 1
+ 1 1 1
1 1 1 0

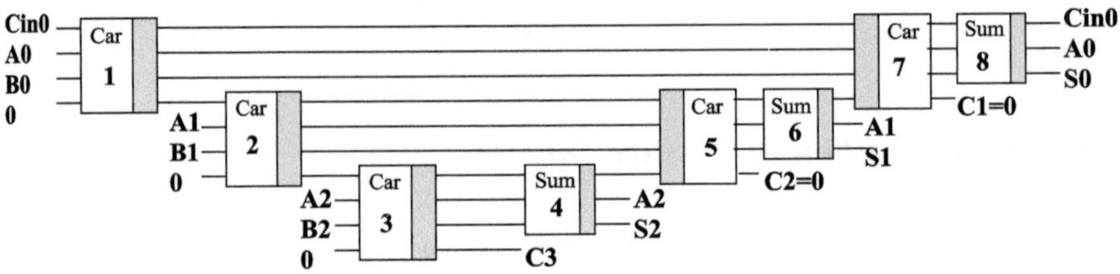

Fig. 7.9 Reversible adder for two 3-bit numbers

input bits are c_{in0}, which is the carry into the operation, and the operands a, b. The carry may be generated as in Fig. 7.8.

For example, if $c_{\text{in}0}$ is zero, while a, b are 1, 1, the inputs on the left are **0, 1, 1, 0**. The carry out is $c_1 = 1$ since $0 + 1 + 1$ equals 0 with $c_1 = 1$. It can be verified by tracing from left to right that the outputs on the right are **0, 1, 1, 1** as required.

Multiple Bit Numbers

Here is an example of adding two unsigned numbers each with three bits. Shown in Table 7.1 is an addend of 7 being added to an augend of 7, or in binary 111 + 111.

The carries into the next places are in parenthesis. These numbers correspond to adding 7 to 7 and arriving at a sum of 14, that is, binary 1110. Figure 7.9 shows how such addition may be done reversibly.

Operations are executed one at a time, from left to right. First the carries are generated; sums are then calculated beginning in the most significant place. The zero inputs to each carry block are "scratchpad" lines for the calculation. As the sums are computed, the carry programming is reversed to restore the scratchpad lines back to zero. Interested readers can trace the result of adding 7 + 7. Let $C_{in0} = 0$; $A0, A1, A2 = 1, 1, 1$; $B0, B1, B2 = 1, 1, 1$. The result obtained in this test case is $C_3, S_2, S_1, S_0 = 1, 1, 1, 0$.

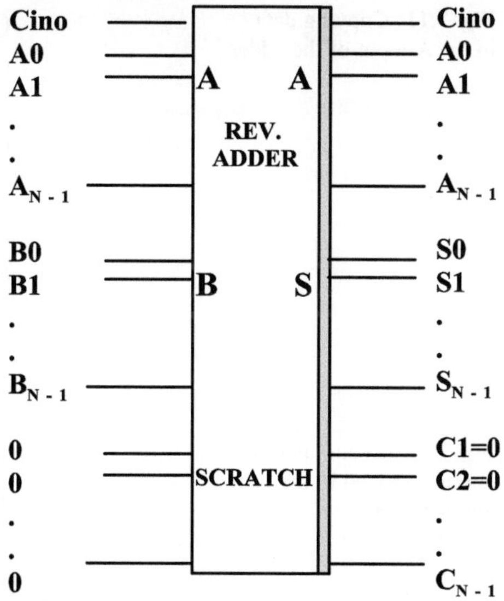

Fig. 7.10 Symbol for reversible N-bit adder

The above approach is easily extended to more bits by using more lines. A symbol for N-bit addition (the addition of two binary numbers with N bits each) appears in Fig. 7.10. Input and output lines are reorganized for convenience.

The signals to port A go through unchanged to ensure logical reversibility; the signals to ports A and B are added; the sum appears at port S. The lines into and out of the port labeled Scratch are scratchpad lines that provide work space for the various operations. The carry out (C_{N-1}) is also the most significant bit in the case of unsigned addition. The above adder works from left to right, as indicated by the bar on the right. Usage from right to left reverses the operations.

N Weights, Z Images

In what follows there will be N numbers or weights for each of Z images, labeled from a to z. The weights are identified as $a_{N-1} \ldots a_1 a_0$, $b_{N-1} \ldots b_1 b_0$, and so on as appropriate to $z_{N-1} \ldots z_1 z_0$.

Computing Priority

Initially the A-inputs of an adder are connected to scratchpad toggles which are initially zero as in Fig. 7.11. The weighting factors are copied one at a time to the A-inputs of the adder. The first to be copied are $a_{N-1} \ldots a_1 a_0$. Copying involves toggling respective scratchpad qubits to 1 on the left if the applied a_i is one, $0 \leq i \leq N-1$.

Fig. 7.11 Copying data to the A-input of the adder

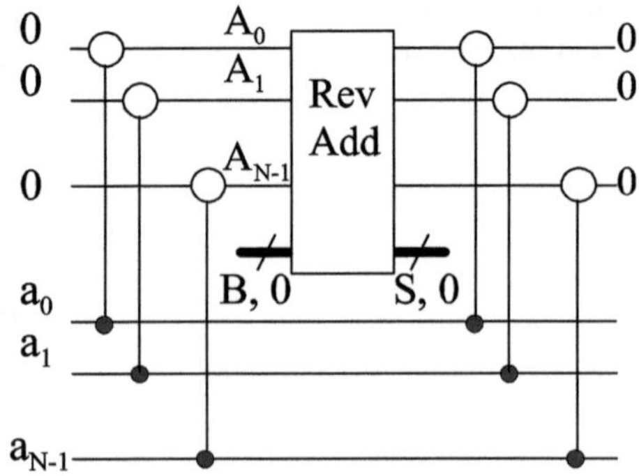

Fig. 7.12 Accumulating another addend to the sum applied to B

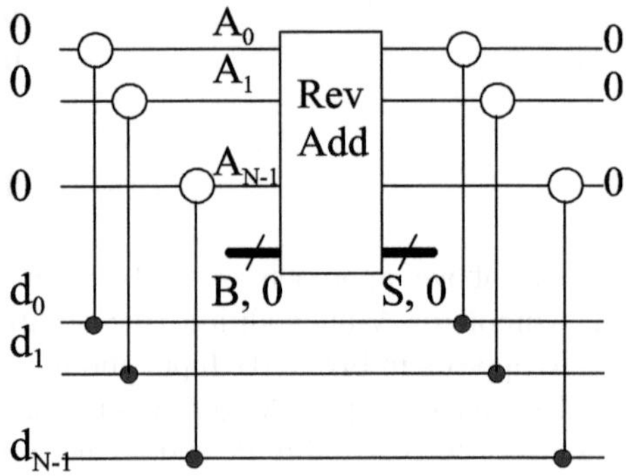

The $b_{N-1} \ldots b_1 b_0$ are held to the B-inputs in the figure so that the reversible adder forms a sum. The qubits holding $b_{N-1} \ldots b_1 b_0$ are replaced with, and are going to hold the accumulation. After the addition, the A-inputs are set to zero by the controlled NOT gates on the right. Note that the $a_N \ldots a_1 a_0$ continue to exist in the simulated qubits where they are stored, so information is not lost.

At this point, the sum consists of the addition of two weighting factors, $a_{N-1} \ldots a_1 a_0$ and $b_{N-1} \ldots b_1 b_0$.

The next set of weighting factors are now copied to the A-inputs. For instance, if $d_{N-1} \ldots d_1 d_0$ is next, an operation as in Fig. 7.12 will be carried out. After the addition, the $d_{N-1} \ldots d_1 d_0$ will continue to store their weighting factors, so information is not lost. Pursuing this method, all weighting factors from all important attributes can be added to calculate a priority for a given image register.

Several Adders Are Required

The priority calculations can be diagrammed as in Fig. 7.13. To begin, weighting factors a and b are added to produce S_1. Then d is added to produce S_2, and so on until the last z_P is added to produce S_{P-1}, which is the net priority.

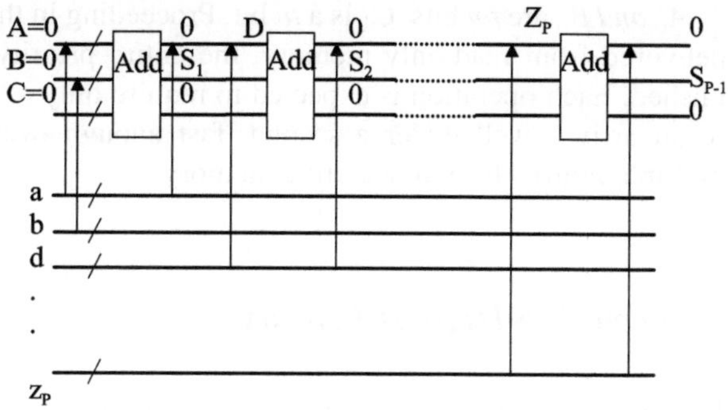

Fig. 7.13 Overview of priority calculations

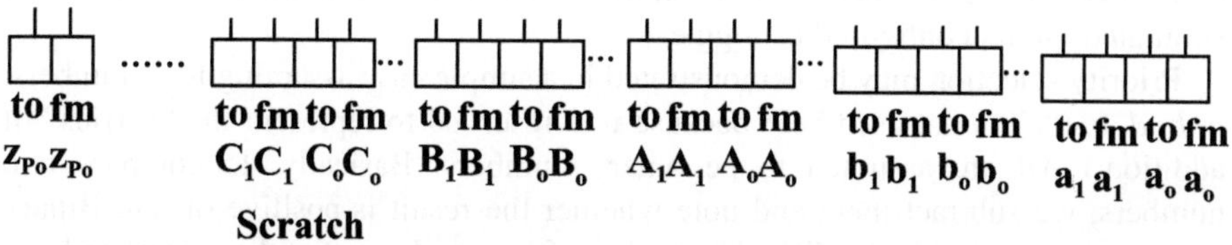

Fig. 7.14 Possible layout for code words

Except for the B-lines, the lines in this figure represent registers of weighting factors, each N bits. The number of weighting factors to be added for each returned image is designated to be P, which requires $P-1$ reversible adders to arrive at a grand sum, the desired priority value. The B-line is going to carry the accumulation, so its bus must have *about* $N + P$ lines.

The wiring diagrams show what must be accomplished by the programming code taken from long-term memory. Space does not permit detailed code, nor do you want to see it, but its main features are described below.

Programming for Additions

Each neural toggle qubit can be controlled by two fields, the *to* and the *fm*. Code words flow from long-term memory as described in a previous chapter. Each simulated qubit has its own *to* and *fm*. Imagine a code word with *to* and *fm* connections to each toggle switch, as in Fig. 7.14.

Each *to–fm* combination is a nano-operation. The first code will be to transfer $a_0 \rightarrow A_0$. The a-toggles are on the right; the A-toggles are in the middle. The code for the first bit a_0 is 00 … 0000 … 0000 … 0010 … 0000 … 0001.

a_0 is the *fm* bit; A_0 is the *to* bit. At some point appropriate inputs will have been applied to the adder, and it is time to add. To compute the carry out of $A_0 + B_0$, for instance, which is $A_0\, B_0$, and to place it in a scratch qubit C_1, that is, $A_0\, B_0 \rightarrow C_1$, apply the code 00 … 1000 … 0001 … 0001 … 0000 … 0000.

A_o *and* B_o are *fm* bits; C_1 is a *to* bit. Proceeding in this way, using such nanocode delivered from read-only memory, the entire priority calculation can be accomplished. Each operation is expected to require only a few milliseconds. So the end result arrives well under a second, fast enough even for adverse conditions of multiple returns from associative memory.

Determining Highest Priority

Integers are available, each a priority for a returned image (a candidate recall). The problem at hand is to find the highest priority, and to use it to enable its corresponding image to enter into consciousness. Rather than random logic as delineated in a past chapter, it is desired to use the structured capabilities of simulated qubits configured as toggles.

Priority selection may be demonstrated in a simple case assuming four numbers, *a*, *b*, *d*, *e*. ("*c*" is not used here because it is reserved to represent the "carries" of addition.) All are assumed to be binary numbers. Basically, to compare two numbers, we subtract them and note whether the result is positive or not. Binary subtraction may be accomplished by taking a 2's complement of the subtrahend and then adding this to the minuend.

Consider $b - a$. The 2's complement of a can be denoted as $a' + 1$. This symbolism means to complement each bit of a and then to add 1. Then add: $b + a' + 1$. The carry out is one whenever $b \geq a$. If $b < a$ the carry out is 0. For example, perform $b - a$ for $b = 4$ and $a = 3$. In binary, $b = 100$ and $a = 011$. The 2's complement of a is 101. Adding 100 and 101 results in 001 with a carry out of 1. Therefore, since the carry out is one,

$$b \geq a.$$

Controlled toggles can determine whether or not $b \geq a$ as in Fig. 7.15, an example kept simpler by assuming a precision N of 3 bits.

The priorities *a*, *b* reside in the registers holding the subject images. For priority a the 2's complement is taken of $a_2\, a_1\, a_o$ as shown in the top left of the figure, which complements bitwise the individual bits going to the A-input. For instance, if a_o is true, a toggle is applied to A_o making it a false. The carry in, C_{in0}, serves to add 1, effectively producing a 2's complement for the A-input (and output) of the reversible adder.

The signals for image "b" are applied to the B-inputs of the reversible adder. Effectively subtraction is done; the carry out is true if and only if $b \geq a$. All the other output information is irrelevant. This implies simplifications in the adder circuit that are of no concern at this point.

A similar circuit for $d - e$ can be used to determine whether or not $d \geq e$.

Next it must be decided which is the largest among *a*, *b*, *d*, *e* by subtracting the larger of *a*, *b* from the larger of *d*, *e*. This parallels the method of a past chapter. The

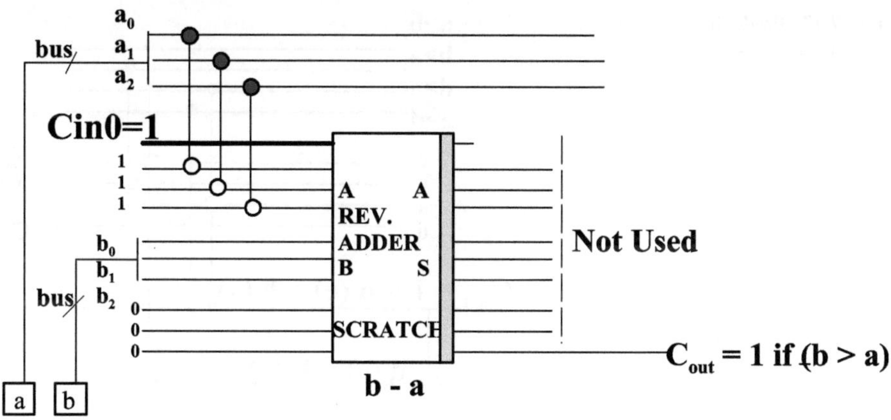

Fig. 7.15 Subtractor to determine if $b \geq a$

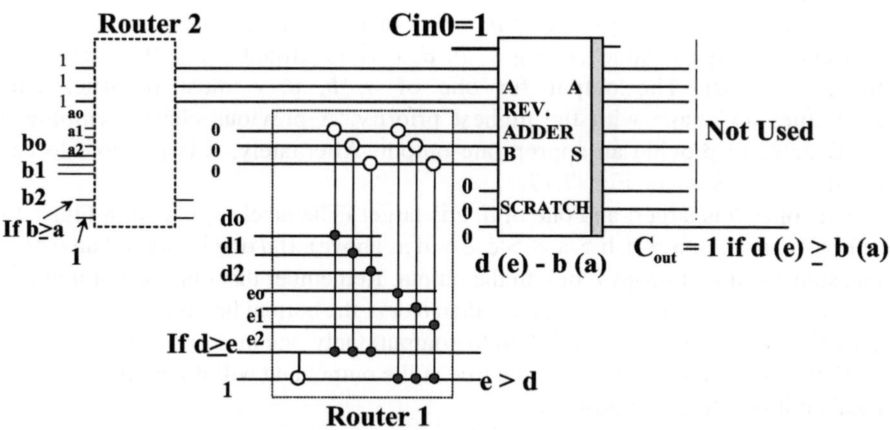

Fig. 7.16 Control circuit to determine the greatest priority

meaning of the expression $d(e) - b(a)$ is that the larger of a, b is subtracted from the larger of d, e. This operation requires a control circuit as in Fig. 7.16.

A neural *router*, router 1, is defined that will apply the larger of d, e. If $d \geq e$, its true C_{out} is applied to a *double controlled NOT* to copy d_2, d_1, d_0 to the B-input of a reversible adder. But if $C_{out} = 0$ a toggle is employed to indicate that $e > d$ at the bottom of the figure. This is applied to a *double controlled NOT* to copy e_2, e_1, e_0 to the B input as shown in the bottom portion of the figure.

Similarly if $a \geq$ b, its $C_{out} = 1$ is applied to a double controlled NOT to copy a_2, a_1, a_0 to the A-input of the reversible adder. But if $C_{out} = 0$ a NOT toggle device is employed to indicate the $b > a$ at the bottom of the figure. This is applied to a *double controlled NOT* to copy b_2, b_1, b_0 to the A-input as shown in the upper portion of the figure. This time, a 2's complement is taken of a_2, a_1, a_0 or b_2, b_1, b_0. This is accomplished by applying logical ones to each input line for the A-inputs,

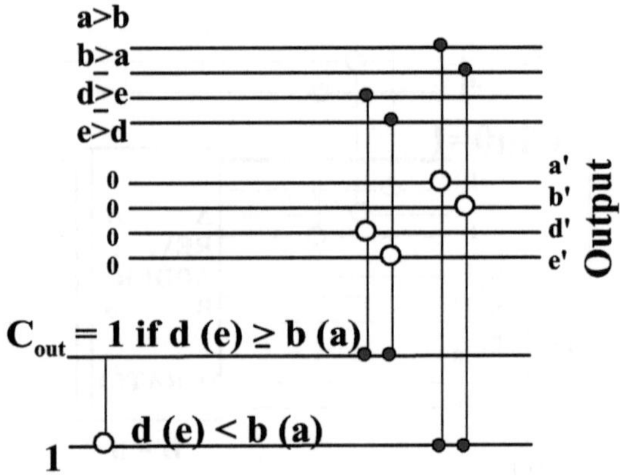

Fig. 7.17 Producing decoded outputs

and reversing each to 0 if the corresponding bit is 1. As before $C_{in0} = 1$ to give a 2's complement.

Note for $b \geq a$ or $d \geq e$, that if $a = b$ or that $d = e$, the system copies b or d.

Hence a largest value among a, b, d, e is identified using the carry outs for $d(e) - b(a)$. The output for one of a, b, d, e must be true. This indentifies an image with the highest priority. A previous chapter employed AND gates to provide an appropriate output. Alternately, it is possible to use toggled signals as in Fig. 7.17.

One of a, b is larger; and one of d, e is larger (The labels a, b, c, d in Fig. 7.17 refer to signals for a > b, b > a, d > e, e > d as above). If d or e is larger than b or a, this signal is used to copy d or e to the output. Remember that only one of d or e is true. On the other hand if d or e is less than b or a, the bottom line is maintained true. This signal is used to copy a or b to the output. Only one of a or b is true.

Finally one of $\boldsymbol{a}'$, $\boldsymbol{b}'$, $\boldsymbol{d}'$, $\boldsymbol{e}'$ will be true at the output and will direct its associated recalled image to consciousness.

Amazing Mental Calculations

Savants known as "mental calculators" occasionally excel at rapid multiplications, divisions, roots, powers, and prime number recognition and other calculating skills. Many savants demonstrate calendar skills. They enjoy asking people for the date of their birth, and then telling them on which day of the week they were born. The most common explanation given for their abilities is that savants have apparent increased ability to remember. While this possibility can account for some savant skills, it fails to explain others.

Savants, for example, often can identify large prime numbers [3]. A person might subconsciously solve the problem of identifying a prime number using parallel processing. Each register of toggles in an array of registers may be initialized to hold the large number n in question, as well as a unique integer m.

Divide the number *n* by integers *m* (beginning from 2 up to the square root of the number) in a search for exact divisors. All division can be in parallel. One or more of the remainders would be zero for a non-prime number; so one can know that a given number is not a prime. But if all remainders are finite, one can know it is a prime number.

This routine can be implemented more efficiently if a complete list of primes up to the square root of the number is available and is loaded into its respective register. Then trial divisions need to be checked only for those m that are prime. For example, to check the primality of 37, only three divisions are necessary ($m = 2$, 3, and 5), given that 4 and 6 are non-prime (Wikipedia, http://en.wikipedia.org/wiki/Prime_number, accessed 2012).

Other methods for prime identification are rather complex for a savant to reasonably learn, and many are unsuitable for a parallel algorithm. Also, based on electroencephalography the mind is limited to about 40 images per second, so most serial algorithms would take too long.

Many calculating tasks can be visualized as a massively parallel set of operations followed by accumulation. For example, when multiplying two large numbers, the multiplicand may be multiplied by each digit of the multiplier in parallel. Then all that remains is to sum the partial products by adding them to an accumulation one at a time to determine the product. Multiplying 678 by 345, for example, can be accomplished by multiplying 678 by 3, while concurrently multiplying 678 by 4, while concurrently multiplying 678 by 5; the product is simply the appropriate sum of the partial products, taken with controlled toggles.

It is interesting to note that the above priority-comparator architecture might be adapted to perform mental arithmetic. For instance, instead of priority, partial products could be determined (in parallel) using codes from long-term memory to activate controlled toggles in a logical register. Instead of a magnitude comparator, an accumulator could be implemented. Thus the above system is capable, at least in theory, of more than priority calculations.

Conclusions

Simulated qubits that operate subliminally perform the necessary digital arithmetic by using routines in the form of codes taken from long-term memory. Such codes flow without conscious exertion, and provide instructions that achieve calculations, in this case integer additions to obtain a priority value, and integer subtractions to accomplish priority comparison and selection.

The method described involves a sequence of operations in which *fm* toggles, if all true, are made to cause *to* toggles to change states. Operations like this may sum weights to obtain the priority for a given image. Similar operations using *fm–to* operations may subtract priorities, to determine which is larger. Finally an image multiplexer is assumed to gate the highest priority image into conscious short-term memory.

Such operations are relatively fast since they are highly parallel. That is, the priorities of each returned image are computed in parallel. Also, when comparing priorities, there is considerable parallelism because pairs of priority values may be subtracted in parallel. This system appears to be compatible with the need for a quick response when danger is present.

References

1. Barenco A, Bennett CH, Richard C, DiVincenzo DP, Margolus N, Shor P, Sleator T, Smolin JA et al (1995) Elementary gates for quantum computation. Phys Rev A 52(5):3457–3467, arXiv: quant-ph/9503016
2. Fredkin E, Toffoli T (1982) Conservative logic. Int J Theor Phys 21(3):219–253
3. Sacks O (1985) The man who mistook his wife for a hat. Summit Books, New York, pp 185–203

Self-Study Exercises

(If you desire to keep a record of your efforts, please show a couple of logical steps leading to the answers.)

Wiring Diagrams

1. Draw a wiring diagram using three lines and assume, from left to right, the following operations: UN on line 1 (bottom), SCN (from line 2 to line 1), DCN (from lines 1, 2 to line 3). Calculate the output on the right when the following is applied to the left:
   (a) 1, 1, 1 (ANS: 0, 1, 1)
   (b) Demonstrate that the operations are logically reversible by applying the answer from part (a) to the right of the diagram to obtain on the left the original 1, 1, 1.
   (c) Assuming these particular UN, SCN, DCN what are the results for the following inputs on the left? Diagram a system of parallel processing. (HINT: Place these numbers into three registers.)

   1, 0, 1;
   0, 1, 1;
   0, 0, 1.

2. Demonstrate that addition and carry are reversible using the following numbers in circuit elements given in this chapter. First run them through from left to right, and then repeat, running the result back through from right to left.

   (a) $a = 1, b = 0, c = 1$

Reversible Addition

3. Refer to Fig. 7.9:

   (a) Trace the calculation at the outputs of each operation using as input 011 and 110.

4. To accumulate numbers as in Fig. 7.13, the number of lines for the accumulation was said to be about $N + P$. Justify this statement.

Coding

5. Modify the code in the form 00 ... 0000 ... 0000 ... 0000 ... 0000 ... 0000 to move a_0 to z_{p0}. (ANS: 10 ... 0000 ... 0000 ... 0000 ... 0000 ... 0001)

Priority Comparison

6. Perform $b - a$ in binary and inspect the carry out to see if $b \geq a$ or if $b < a$:

   (a) For $b = 5$ and $a = 5$
   (b) For $b = 7$ and $a = 1$
   (c) For $b = 2$ and $a = 7$
   (d) For $b = 7$ and $a = 0$

7. In Fig. 7.16:

   (a) What is applied to the B-input when $d = 011$ and $e = 100$? (ANS: The larger of the two)
   (b) What is applied to the A-input when $a = 110$ and $b = 011$?

8. In Fig. 7.17, assume $d < b$, but that $b \geq a$ and $d \geq e$.

   (a) What are the truth values at a′, b′, c′, d′? (ANS: 0, 1, 0, 0)
   (b) Can more than one of a′, b′, c′, d′ go true? (ANS: NO)
   (c) If priorities b and a are equal, which corresponding image is permitted into conscious short-term memory? (ANS: b)

Amazing Mental Calculations

9. Identify a fast pencil and paper way to compute 123,456 × 654,321.
10. Identify a fast pencil and paper way to obtain an approximation (to the nearest integer) for the square root of 150.
11. Identify a fast pencil and paper way to determine whether or not 101 is prime.

Chapter 8
Long-Term Memory Neural Circuits, Fast and Precise

Introduction

Long-term memory has a great many logical properties, so it follows that its neurons most likely possess a logical structure. Perhaps the most obvious logical property is associability, since fragments in consciousness result in associative recalls from long-term memory. Fragments in consciousness refer to collections of attributes, that is, basic colors, shades, shapes, edges, and other components of a mental image.

Cues are selected from conscious short-term memory (STM) and sent "down" into long-term subconscious associative memory; images with attributes that exactly match all cues are returned for possible updating of conscious STM. It is emphasized that images are called forth only for exact matches. Gratefully, old memories are dormant until receiving an exact set of cues since serious overload would result if we had to see all returns, even those totally irrelevant.

A neuron never has to multiplex two or more attributes, and so will not have to endure incessant jumping between unrelated signals in the same axon. Multiplexing is excluded in the model of this book partly because neurons are plentiful and can be dedicated to given attributes, and partly because dedicated neurons are necessary in order to define a plausible neural system.

It follows that attributes that have their own dedicated column of neurons and are distributed. In the model below, neural clusters are dedicated to a given attribute; each attribute connects to specific locations in a *word* of *STM* and to corresponding locations in words of *long-term* memory. Location is thought to define the meanings of attributes. Locations obviously exist physically, although we do not yet know exactly where.

For example, the color fire engine red might be the 37th neuron from the left in the word of STM; it is also the 37th neuron from the left in each word of long-term memory. Of course, this 37th neuron will be active only when red is needed, such as when recalling a red fire engine.

J.R. Burger, *Brain Theory From A Circuits And Systems Perspective*, Springer Series in Cognitive and Neural Systems 6, DOI 10.1007/978-1-4614-6412-9_8,
© Springer Science+Business Media New York 2013

Although irrelevant to a logical analysis, one may wonder exactly where individual memories are located, with the idea in mind that memories might be altered by surgery. In reality, there are redundant neurons spatially distributed to encode a given attribute, to increase system reliability. And there is duplication in the left and right hemispheres. Memories of attributes are duplicated in several places, so it is unlikely that a memory can be surgically removed.

On the other hand, redundancy is strictly limited, since if it were not, a brain would be quite inefficient, and burdened with an excessive number of neurons for exactly the same purpose. In order to discuss memory without loss of generality, the model below assumes one memory element per attribute.

It is assumed below that any arbitrary subset of attributes taken from STM can be used as cues for an associative search. For example, cues may be fully identical to the attributes in conscious STM or cues may be any subset of such attributes, or in an extreme case, a cue may be a single but crucial attribute. The nature of attributes does not affect the analysis. For example, attributes for emotions are just ordinary attributes as far as memory is concerned, although once recalled they may have special potency.

Memorization is a system property that is modeled as working hand-in-hand with associative recall. A particular logical structure for neural memory suitable for recalling and memorizing is derived below.

Words of Memory Elements

To review, memory elements are (1) recursive (multivibrating) neurons or (2) neurons easier to trigger because of long-term potentiation. Variations, probably combinations of the above models as discussed in a previous chapter, are taken in this book to create a basic element of long-term memory such that when needed, provides an attribute signal, but only if that attribute is part of a memorized image.

A single long word is not viable for long-term memory since it would be inefficient; sub-images would have to be strung end to end making them tough to retrieve. Also, words might end up with differing random lengths, making them difficult to place into conscious STM. For efficiency, this chapter assumes words of approximately equal lengths so that all may be addressed in parallel, associatively.

Standard Memory Cells

Long-term memory can be modeled as a connection of *standard memory cells* illustrated in Fig. 8.1. A standard cell in circuit parlance is a particular circuit that is designed to connect easily to other identical circuits [1, 2]. The term has nothing at all to do with a biological cell. The Q in this figure represents the output of an

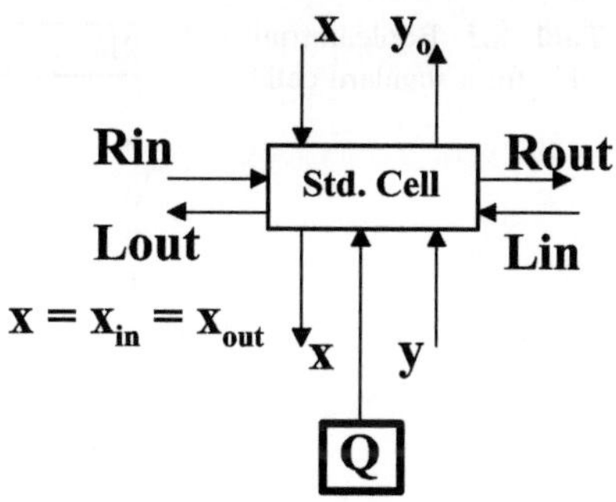

Fig. 8.1 Standard memory cell for biological memory modeling

Table 8.1 Definition of memory cell signals

Q	Memory signal from memory neuron
x	Cue for given feature (input)
$x_o = x$	Cue to cells below (output)
y	Memory from cells below (input)
$y_o = y$	Memory of a given feature (output)
Rin	Right-going local bus signal (input)
Rout	Right-going local bus signal (output)
Lin	Left-going local bus signal (input)
Lout	Left-going local bus signal (output)

Table 8.2 Desired logic for standard cell

	Conditions	Rout
1	Rin = False x either True or False Q either True or False	Rout = Rin = False
2	Rin = True x = False, no feature Q either True or False	Rout = True
3	Rin = True x = True Q = False	Rout = False
4	Rin = True x = True Q = True	Rout = True

element of long-term memory. An enable signal E, not shown, is available to activate cells and memory elements as needed.

A standard memory cell may be synthesized based on input and output behavior as defined below. Signals are defined as in Table 8.1.

Note that neurons sometimes work in pairs: Cue signals from a cue editor system, the x signals, travel down. Recalled features, the y signals, travel up and into a recall referee system. Memory word bus signals also work in pairs: A right-going bus signal travels to the right, from Rin to Rout along a local bus. A left-going bus signal travels to the left, from Lin to Lout. Rin and Lin are going to be energized by outside *enable* signals. The desired logic has a *truth table* (Table 8.2).

Table 8.3 Boolean truth table for a standard cell

Rin	x	Q	Rout
0	0	0	0
0	0	1	0
0	1	0	0
0	1	1	0
1	0	0	1
1	0	1	1
1	1	0	0
1	1	1	1

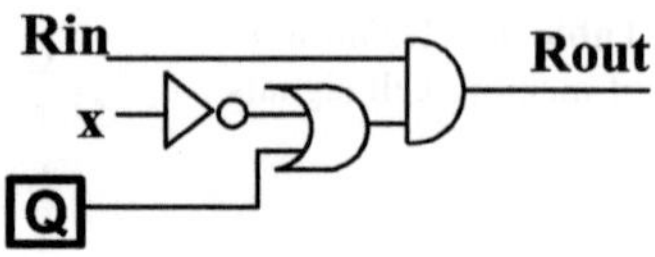

Fig. 8.2 Rout–Rin logic

In Boolean algebraic terms False is 0; True is 1. The *first* entry means that if there is no bus signal coming from the left (Rin = 0), then none is transmitted to the right (Rout = 0). This keeps a memory word dormant until needed. The *second* entry in the table means that if there is no cue ($x = 0$) then there is no change in Rout and Rout = Rin = 1. The *third* entry means that if there is a cue for an attribute, or $x = 1$, but there is no memory of this feature in this word, or $Q = 0$, then bus operations terminate (Rout = 0). The *fourth* entry means that if there is a cue for an attribute ($x = 1$) and if there is also a memory of this attribute ($Q = 1$) then there is no change in Rout, and Rout = Rin = 1. A proper binary "truth" table can now be constructed (Table 8.3).

In this table, entries 1–4 express condition 1, no input (Rin = 0) gives no output (Rout = 0). Entries 5, 6 express condition 2, no cue permits an output, with or without memory. It permits a recovery of attributes that are not covered by the cues. Entry 7 shows condition 3, cue but no memory gives no output, which will prevent a return from this word. Entry 8 shows condition 4, both cue and a memory of the attribute represented by the cue give an output.

The appropriate Boolean equation for Rout, derived from the truth table using standard symbols (•, +, ′ for AND, OR, NOT) is

$$\text{Rout} = \text{Rin} \bullet [x' + Q].$$

x' represents a NOT function of x. A logic circuit corresponding to this equation appears in Fig. 8.2.

This can be implemented using neurons. A similar neural logic circuit may be written for Lout in terms of Lin (not shown).

Fig. 8.3 Readout logic

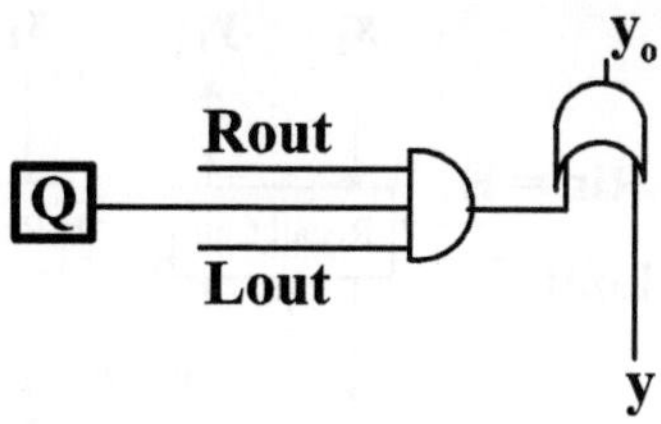

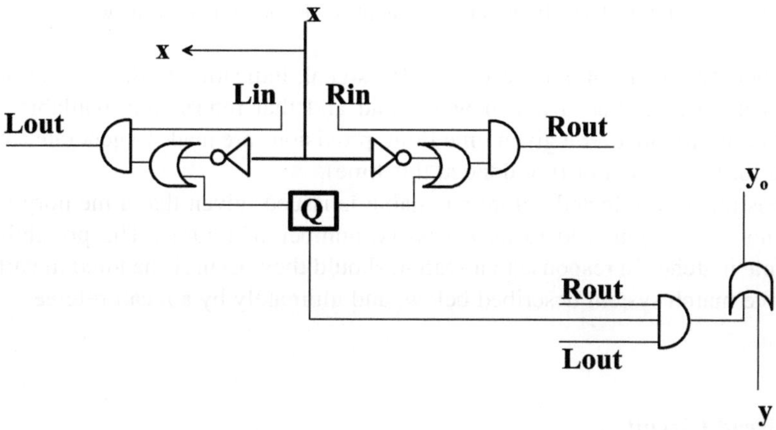

Fig. 8.4 Logic for a standard cell of long-term memory

Readout Details

The output of a standard cell is y_o. This output will be true only if both Rout and Lout are true for a given cell and also if Q is true. Requiring a true value for Rout and Lout, as well as Q prevents false recall signals. The standard cell is organized below in such a way that Rout and Lout cannot both be true unless there is output from each cell in the word. Figure 8.3 illustrates an appropriate readout circuit.

The neural logic in a standard cell can now be summarized as in Fig. 8.4.

The Rout and Lout signals may be connected to give a word with any number of standard cells. An example of four cells is shown in Fig. 8.5.

The Rin signal will take a certain (short) amount of time to ripple across the memory word; so will the Lin signal. They meet in the middle. Where the signals overlap, output is possible since the local Rout and Lout must be true in order to have output. Thus output begins from the middle of the word and soon is available at both ends.

Note that the x_i inputs ($i = 1, 2, 3, 4$) coming from above are also the $x_i = x_{oi}$ outputs. E represents a memory system enable that energizes Rin entering on the left and Lin entering on the right, and possibly also prepares the memory elements to be used. When E goes true it begins a memory search.

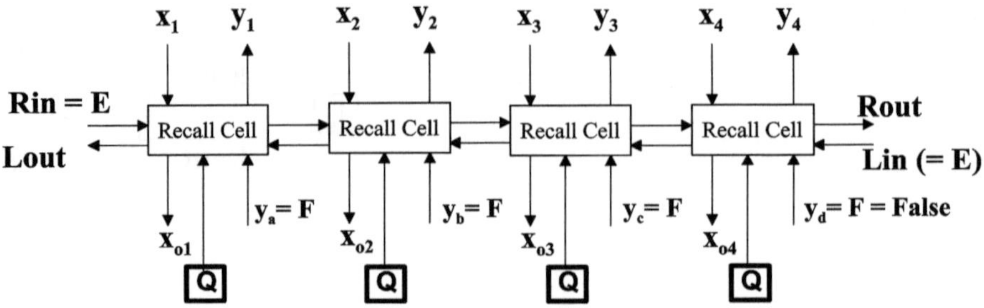

Fig. 8.5 Circuit for reading attributes in an example of a word of long-term memory

Either Rout or Lout may serve as a *Hit* signal, indicating to the cue editor, and the recall referee, that a match was found and that returns are available. If no matches are found after a given time limit, provisions are made to generate a *NoHit* signal for the cue editor (not done at this time).

Cues must be selected before an enable is issued, given that a memory search with no cues might lead to an excessive number of returns. The possibility of multiple matches in response to a search, should they occur, is handled in part by a multiple match system described below, and ultimately by a recall referee.

Multiread Circuit

When an inadequate number of cues are presented to an associative memory, one normally expects multiple matches to the cues and thus multiple returns. This requires special handling mainly because two images cannot simultaneously occupy a return bus.

Intuitively a person cannot picture two images exactly at once. As an experiment in multiple recalls, try to think of two things at once, such as your car and your kitchen sink. It might be noticed that fuzzy images seem to alternate with each other, although in rapid succession.

Neural logic prevents totally separate images from being recalled at exactly the same instant. In the above figure with four standard cells, the *y* inputs are coming up from other parts of memory. They were assumed to be false and will be false as long as there are no other matches to the given cues. If there are other matches to the given cues in the circuits below, it is necessary to delay them.

Figure 8.6 shows how. A match between cues and stored memory words means that Rout is true. A complement of Rout is sent to the inputs of AND gates on each *y* signal coming up from below. This will lock out all other matches to memories below the top most return, at least until the Rout and Lout signals clear.

Figure 8.7 shows how a word is disabled once it produces a return. A simulated qubit toggle is initialized to true via a multi-match initialization signal or *MMINIT*, but should Rout indicate a match, the toggle gets set to false as the word is read.

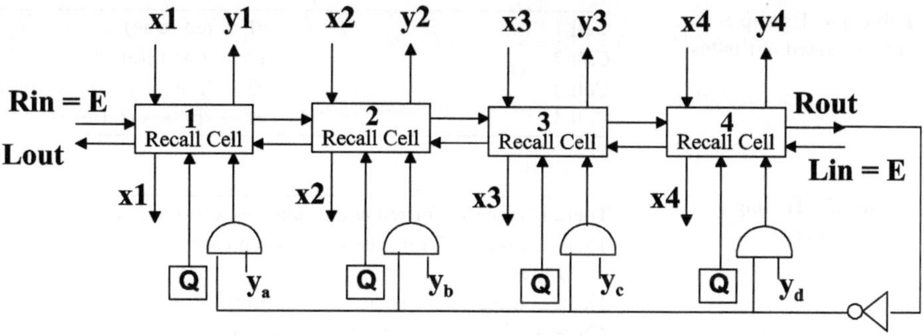

Fig. 8.6 Multiple match resolution

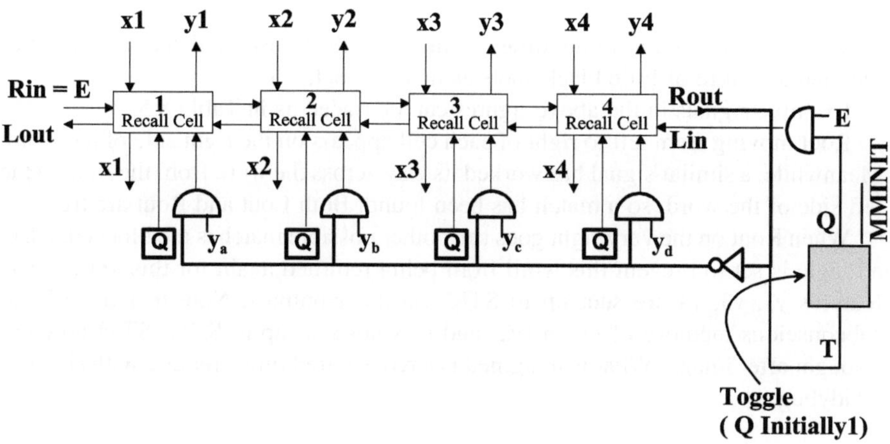

Fig. 8.7 Preventing returns of the same image

The toggle remains false even when Rout subsides to its rest value. This prevents it from being read again and permits other words to provide a return.

A *cycle* of searching involves reapplying the enable *E* timed to get additional returns for consideration, while keeping cues unchanged. After a cycle, new cue combinations from the cue editor will find new returns in a memory search, since cues have an exacting nature.

Note that *E* and *MMINIT* are instigated by the *New* signal previously defined as signaling a new set of cues for a new memory search. Appropriate signals are easily produced by neural networks in the styles previously covered.

Recall Example

Assume a list of attributes as follows: Red, Green, Little Black Spots, Legs. In attempting to recall the details of a particular little creature, a cue editor provides

Table 8.4 Example of memorized attributes

Cell 1	$x1$ = red (true)
Call 2	$x2$ = green (false)
Cell 3	$x3$ = little black spots (true)
Cell 4	$x4$ = six legs (false)

Table 8.5 Tracing a memory search

There is a cue $x1$ for red and a subconscious memory of red
There is no cue $x2$ for green and also no subconscious memory of green
There is a cue $x3$ for spots and a subconscious memory of spots
There is a no cue $x4$ for legs but there is a subconscious memory of legs

some but not all features of the little creature. Say the transmitted cues are the color red, and a texture of little black spots as in Table 8.4.

Example signals in the above figures can be traced as in Table 8.5.

Rout moving from left to right of each cell appears on the right side of the word. Meanwhile, a similar signal has worked its way across the word from the right to the left side of the word, so a match has been found. Both Lout and Rout are true.

When Rout on the very right goes true, other possible matches are blocked. Also, a toggle is set to prevent this word from being returned again for this search. The features $y1$, $y3$, $y4$ are sent up to STM via the y-outputs. Note that there is no subconscious memory of green ($y2$) and it is not sent up to STM. STM now has a sought after image. What is imagined is a red, spotted little creature with six legs, a ladybug.

Models for Memorizing

A memory model must incorporate the intuitive knowledge that memories are formed practically instantly, and cannot be erased easily, nor can they be overwritten. New memories require available memory, referred to as *blank words* of memory. So it is reasonable to assume that blank memory elements are available, ready to be used, since otherwise new memories would be rare.

Memorizing is not the same as learning. Learning involves neural redirection and synaptic formation, which takes time. Learning new sequences, for example, may involve inteneurons and synapses that go from one word of long-term memory to another in a controlled way without conscious involvement. Redirection and growth take rehearsal time.

In contrast, memorization of details seems to occur practically immediately, as is common experience during emotional events such as shocking deaths, horrible losses of life and property, and intense moments in relationships. Beyond this, it is known that those gifted with photographic memory instantly accomplish massive

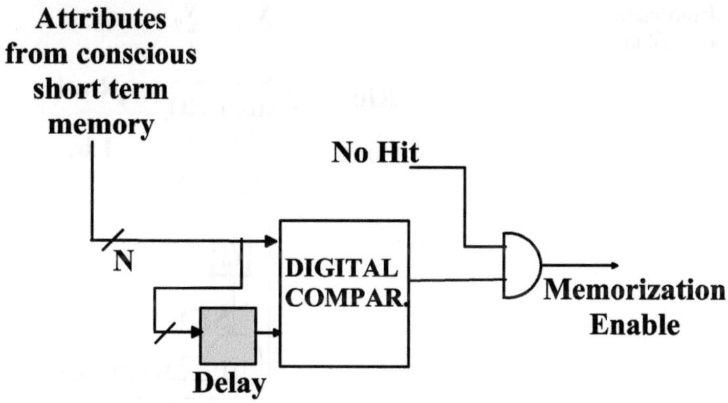

Fig. 8.8 Memorization enable circuit example for two rehearsals

amounts of memorization, books full, that is apparently permanent. Therefore it is assumed below that memorization is instant and permanent.

A blank word must have access to all possible attributes since, as is common experience anything can be memorized. To reach each and every attribute, the same circuits as for search and recall are assumed used for memorization.

Once a memory is formed, a vast number of possible attributes are unused, and never will be used in conjunction with a given word and its image. Because of a phenomenon known as neural plasticity unused neurons will reconstitute themselves as part of some other function. Perhaps they will serve in new blank words. This suggests that words of long-term memory will be sparse, retaining only essential attributes.

Memorization Enable

Memorization in the above system is triggered by a signal termed *Memorization Enable* which is sensitive to recurring images in conscious STM, under specific conditions: *Memorization Enable* is forced true with a digital filter whose output goes true for a periodically repeated input image or action in consciousness. This might be accomplished as in the example of Fig. 8.8 which requires two repeats before memorization. At the same time there must be no matches (*NoHit* = *True*) from long-term memory. That is, the image must not already be in memory.

So if a given image is not in long-term memory and appears in conscious STM repeatedly, it will be committed automatically to long-term memory. Note that a digital comparator is a variation on the magnitude comparator discussed under priority comparison in a previous chapter.

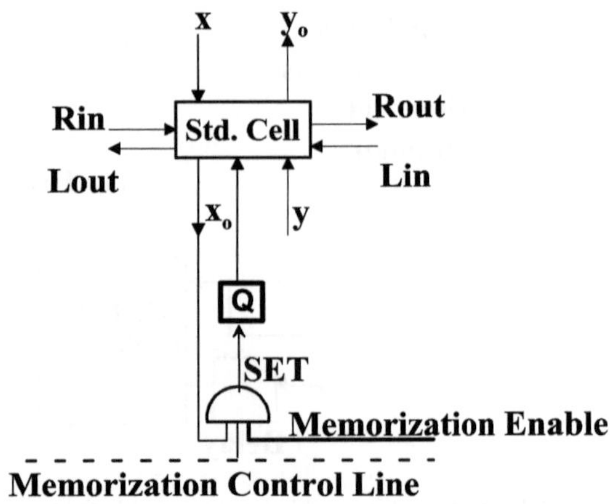

Fig. 8.9 Memorization circuit for an attribute in a word

Circuit Model for Memorizing New Memories

The neurons that convey cues for memory search may also serve as conveyers for attributes to long-term memory. However, now they are not just a sampling of cues but the entire image. Figure 8.9 shows how the cellular recall model can work for memorization.

The circle with the Q represents a blank memory element. Memorization begins when *Memorization Enable* goes true. However, for memorization to occur, a control line termed the *Memorization Control Line* must be true.

Multiwrite Circuits

There have to be controls to select only one blank word so that not all available words are programmed with the same information. This may be accomplished using a neural *multiwrite* which is a neural form of a *one-hot* circuit. Memorization can be regulated by signals to a *Memorization Control Line*, whose purpose is to give multiwrite capability.

Multiwrite Using Long-Term Memory Elements

Figure 8.10 illustrates a simple system based on memory elements, the elementary kind whose outputs are initially false, but which once set, the outputs stay true indefinitely. The XOR at the bottom of the figure has one input held to true, so its output to the first *Memorization Control Line* is true. All other memory elements for control are resting at false, so they are removed from the system.

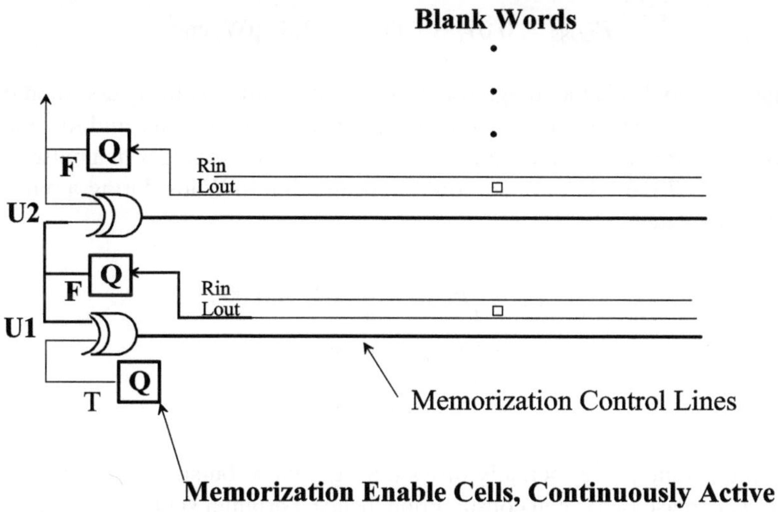

Fig. 8.10 Programming blank words one at a time (multiwrite)

It is not obvious how the bottommost memory element becomes true. It could occur with the first *Memorization Enable* signal. One may imagine clusters of blank words; once one is filled, another cluster is enabled via an interneuron, beginning from the bottom.

But once the associated word is written, its Lout goes true for a moment, thus setting the next memorization enable cell. Now the given XOR has two true input, so the output will be held to false, preventing further changes in the given memory word. The inputs to this XOR need to be timed properly if a dendritic gate is assumed. Otherwise a multilayer XOR made up of enabled OR, AND, NOT will be necessary.

Memory elements must produce outputs and so dissipate a small amount of energy. However, neural pulses are relatively slow, in the millisecond range, so losses in series resistance are minimized. Membrane losses can be shown to actually drop during pulsing, as discussed below. So dissipation is minimal.

Calories for Memorizing

Normally one expects increased quantities of deoxygenated hemoglobin, due to energy consumption during neural pulses, and a *reduction* in the magnetic resonance signal. A noteworthy mystery is that during heightened neural activity, this actually occurs an *increase* in the magnetic resonance signal [3, 4].

The mystery may be explained, at least in part, by membrane conductance loss. Neural membrane conductance is typically estimated to be 0.3 mS/cm^2. Membrane voltage at rest is about −70 mV. So at rest, electrical power density consumed in W/cm^2 in each and every neuron is

$$P_{\text{LOSS}} = 0.07^2 \times 0.0003 = 1.47\,\mu\text{W/cm}^2.$$

Energy in joules is the integration of power over time. A fairly common type of neural pulse can be modeled as equivalent to being high for 1 ms and low for 3 ms, that is, high for ¼ of the time and low for ¾ of the time. Voltages in a pulse train go from about −70 to +40 mV. So power consumption in Watts during a pulse burst may be calculated as

$$\begin{aligned} P_{\text{ACT}} &= \frac{3}{4} P_{\text{LOSS}} + \frac{1}{4} 0.04^2 \times 0.0003 \\ &= \frac{3}{4} 1.47 + \frac{1}{4} 0.48 \\ &= 1.222\ \mu\text{W/cm}^2 \end{aligned}$$

So membrane power actually drops during a pulse burst. A decrease in power means a decrease in oxygen consumption, hence a brighter spot in an fMRI, all else being equal. The contrast is $1.47 - 1.22 \approx 0.25\ \mu\text{W/cm}^2$ or 16.87 % of resting power density, about what is observed. This could be a simple explanation of why during heightened neural activity, there actually occurs an *increase* in the magnetic resonance signal.

Alternative Multiwrite Using Toggle Elements

Each blank word may be equipped with a *Long-Term memory Toggle* all of which are initially set to false or rest. But the first toggle for the first blank word must have its toggle set true for a brief time. Figure 8.11 shows a circuit that assumes a signal

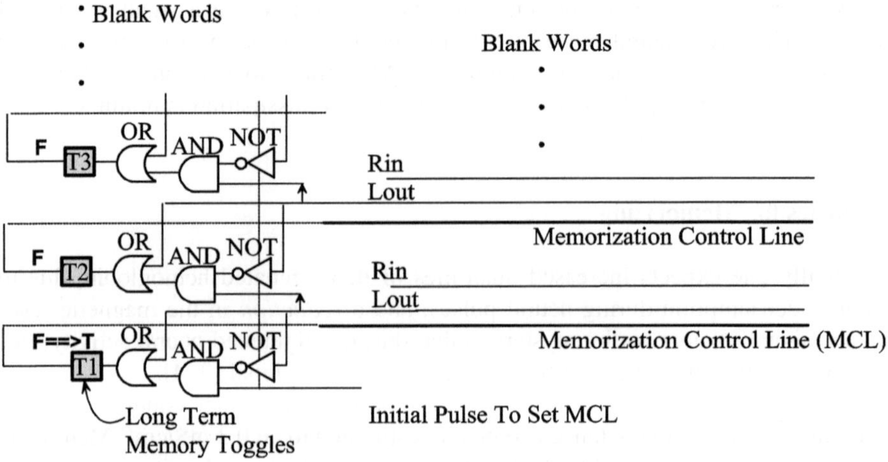

Fig. 8.11 Multiwrite using toggle elements

from past events to set the first toggle T1 to true and to begin the filling of a given cluster of blank words.

The *Memorization Enable* serves to enable the NOT gates shown. If the blank word above a given blank word is at rest, as they all are initially, the AND causes a toggle in the subject blank word from false to true. This true is applied to the subject *Memorization Control Line*, which is made to cause an image to be written into a blank word. Since what is being written is still applied, there will be a Lout signal. This serves to toggle T1 back to false and to toggle the next T2 to true. So the next blank word is now ready to receive an image as soon as *Memorization Enable* goes true again.

Except for the first blank word, the others depend on Lout to modify their toggles. Consequently they are not all toggled at once, only one at a time. For all except one, their state is false, or at rest, so this approach is efficient from the point of view of minimizing neural pulsing.

Memorization Versus Learning

Simulated Qubits in Savant Memorization

Gifted savants sometimes possess photographic memory; their calendar skills require memory; and memory may be needed for exceptional mental calculators. But what does it mean to have great memory? Memory is useless unless there are cues for recall. Memories of random information, for instance, are more likely to be recallable if they are connected together somehow by mnemonics, a form of *encoding*.

What was originally thought to be amazing memorization by savants may in fact be rapid learning. *State machine* learning is defined to be a type of subconscious learning that uses hidden *pointers* to produce a next step based on previous steps. Unfortunately, this type of learning requires rehearsal time. After suitable rehearsal, interneurons become involved and eventually, there is synaptic growth. But savants with memory skills can look over a long text and recite it almost immediately and, we are told, they can recall it years later. Apparently savants require far less rehearsal than average in order to learn sequences.

Learning a Long Sequence

S*tate machine learning* is thought to be responsible for all of the many procedures humans employ unconsciously, including routine actions and recitations without conscious involvement, or contemplating. The big advantage of state machine learning is that mental procedures, which range from the mundane to the advanced, do not have to undergo recall processing through conscious STM each time they are used.

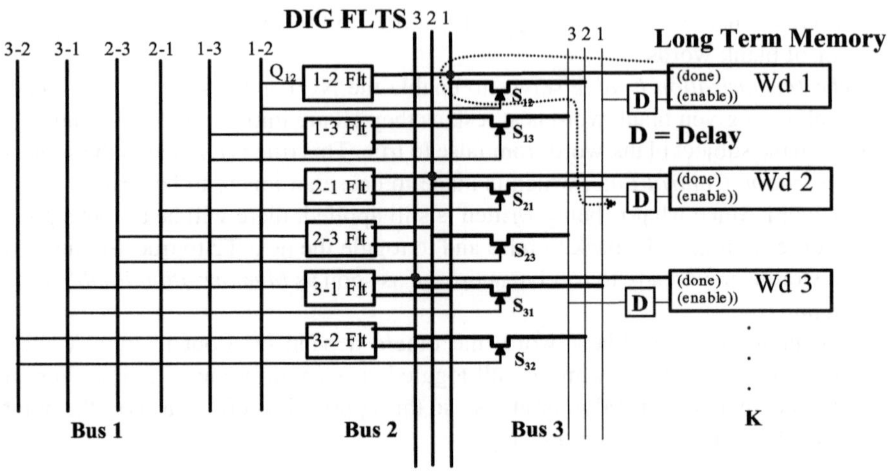

Fig. 8.12 Model for learning; three word state machine within subconscious long-term memory

Figure 8.12 attempts to depicts a small-scale subcircuit that is capable of fast state machine learning, stepping from any given memory word to any other memory word within a given region of subconscious long-term memory. The *Filters* (Flts) in the figure detect repetition and subsequently hold a signal that enables synaptic growth. Synapses (S_{ij}) are modeled as field effect transistors serving as switches in the figure. Interneurons underlie the biological equivalent of this circuit. Interneurons may hold learning connections and release neurotransmitters to foster synaptic growth, thus supporting the S_{ij} in this circuit model for sequential learning.

The design in the figure implies that a system may learn to perform state transitions from any given word to any other given word in a finite block of K words within long-term memory. Figure 8.12 shows three words, although more are possible. The theoretical number of possible sequences is $K!$, which is rather large, but by posing simple restrictions, such as having only ascending sequences, this number is greatly reduced.

For example, consider the sequence 1–2–3 that has been used, say, ten times so that it is well learned. Then, whenever Word 1 is addressed (associatively) by STM, after a short delay, Word 2 is addressed directly by Word 1. The path of the signal is indicated by the dotted line in the figure. Then after another short delay, Word 3 is addressed directly by Word 2. This will accomplish a three-step procedure without conscious involvement involving cue editing and recall refereeing.

Note that switches S_{ij} $(1 \leq i \leq K, 1 \leq j \leq K, i \neq j)$ enable the state machine. For example, switch S_{12} applies a done signal from Word 1 to the delayed enable of Word 2. Switches are implemented biologically with synaptic connections promoted by interneurons carrying serotonin and possibly other synapse-promoting neurotransmitters.

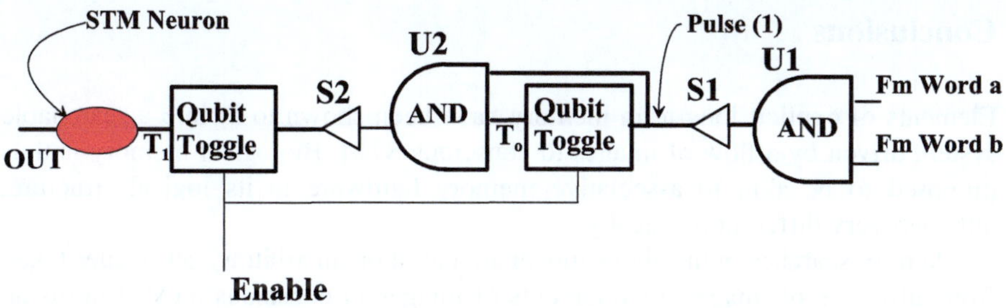

Fig. 8.13 Filter example that detects two repetitions for learning

In general for K words $[K(K-1)]$ timing filters are needed. A sequence in this model may go forward or backward. This implies a like number of bus lines for the filter outputs and a like number of switches S_{ij} $(1 \leq i \leq K, 1 \leq j \leq K, i \neq j)$. Also there must be K lines for memory word outputs and K lines for memory word inputs. The interneuron overhead is high, but probably consistent with the complexity of brain circuitry.

Savant Learning

Synaptic growth for implicit learning takes excessive time and thus accounts poorly for photographic memory which can be formed instantly. Special rehearsal filters may be available to explain rapid learning [5]. Consider a simplified digital filter model for rehearsal as in Fig. 8.13. A toggle pulse occurs when a signal goes from one particular word to another word, thus passing through the input AND gate U1. Only one pulse is passed through a weak synapse (triangular symbol) such that T1 toggles.

T2 cannot toggle because by the time T1 toggles, the input pulse is gone. The AND gate U2 does not permit T1 to toggle until a second occurrence of a signal from one particular word to another word. After two such signals, the output goes true (A similar cascade of toggles was suggested in a previous chapter concerning a recall referee). Thus we have a rehearsal filter example with output after only two rehearsals, causing an automatic move from one memory word to another.

The output of this rehearsal filter holds via a STM neuron; it holds until long-term potentiation and a synapse can be developed. "Memory" savants therefore are proposed to be not memory geniuses, but learning geniuses. They need fewer rehearsals. This would permit faster learning of sequences, resulting in a long memorized recitation without conscious effort.

To avoid having a system attempt to learn too much, it may be necessary to regulate learning using global interneurons; for example, an *Enable* might be activated by a savant's motivation to memorize something interesting. Simulated qubits with rehearsal filters for quick learning in conjunction with quick memorization would explain what is perceived to be phenomenal memory.

Conclusions

Elements of explicit long-term memory have been shown to fit into a searchable system driven by a flow of images to conscious STM. Biological memory is thus proposed to be akin to associative memory hardware in its logical structure, although very different physically.

Memory searches in the above model are based on an arbitrary set of cues taken from attributes of images, or fragments of images in conscious STM. Confusion caused by an episode of multiple returns is prevented by an elementary multi-read system. A simple lockout system permits images to be returned to an attributes bus one at a time. A simulated qubit in the form of a toggle serve to tag an image once it is read, so that it cannot be read again during an episode of multiple returns. Multiple responses are therefore constrained to occur in a sequence, to be presented to a recall referee.

Memorizing is accomplished by assuming available (blank) words that are automatically filled under certain conditions. Conditions are (1) repeated usage of a given image in STM, that is, rehearsal and (2) that the image does not already exist in long-term memory. Writing to more than one blank word at a time is prevented by a basic multiwrite system that is given above. The given multiwrite system is essentially a one-hot state machine implemented with simulated qubits in the forms of toggles, with neural logic to force a filled blank word to point to the next.

Explicit memory, by the way, includes STM and long-term memory. In particular, subconscious explicit long-term memory in humans can be formed instantly, too quickly for synaptic growth, and too fast for long-term potentiation. It seems more likely that the particular attributes of a memory are captured quickly by recursive neurons, and then held as long as necessary until some form of long-term potentiation occurs.

Gifted savants are famous for their powers of memory such as remembering books they have read. Memory is fundamental to savants, but what does memory mean. Pure memory is meaningless unless the connections between the various images of memory, and these must also be remembered. Recalling the contents of a book, which is something savants can do, is proposed to be a manifestation not of fast memorization but of quick learning.

Learning for ordinary people can be modeled as synaptic connections within subconscious long-term memory. Usually connections are formed rather slowly after much rehearsal, as detected by a rehearsal filter. Once learning is in place, synaptic connections define a state machine within subconscious long-term associative memory. State machines, by the way, are logical structures that sequence routine actions, movements or recitations without conscious involvement, or contemplating.

Savant learning may form quickly within subconscious long-term memory; this can be accomplished with sensitive rehearsal filters. Rehearsal filters are proposed to be equivalent to simulated qubits that toggle and count rehearsals. In all likelihood savants both memorize and learn with a lower count of rehearsals. While learning, they develop a state machine held by interneurons that encourage synaptic growth. Once in place, subconscious long-term memory jumps from one word of information to another as fast as the information can be expressed, without having to run each step through conscious STM. In this way, without knowing how, a savant remembers amazing sequences.

References

1. Pagiamtzis K, Sheikholeslami A (2006) Content-addressable memory (CAM) circuits and architectures: a tutorial and survey, IEEE J Solid State Circuits 41(3):712–727
2. Burger JR (2009) Human memory modeled with standard analog and digital circuits: inspiration for man-made computers. Wiley, Hoboken, NJ
3. Huettel SA, Song AW, McCarthy G (2004) Functional magnetic resonance imaging. Siniauer Associates, Inc, Sunderland, MA, pp 162–170
4. Ogawa S, Lee TM, Nayuak AS, Glynn P (1990) Oxygenation-sensitive contrast in magnetic resonance image of rodent brain at high magnetic fields. Magn Reson Med 14(1):68–78
5. Burger JR (2011) Qubits underlie gifted savants. NeuroQuantology 9:351–360

Self-Study Exercises

(If you desire to keep a record of your efforts, please explain your answers in a clear way)

Memory Circuits

1. A neuron may accomplish Boolean logic:
   (a) List major differences between neurons and CMOS (HINT: Pulses vs. voltage levels; dendritic logic with timing vs. enabled logic)
   (b) The XOR is generated by a standard Boolean equation. Show a logic circuit for z that corresponds to this Boolean equation (HINT: You may use five basic gates)

$$z = xy' + x'y.$$

   (c) Modify the XOR circuit to give a non-zero output only if an enable signal E is true (HINT: Use three-input AND gates with one input an enable signal)
   (d) Demonstrate with reference to the XOR equation that it can be used to generate a NOT function of x. Illustrate the NOT using an XOR circuit (HINT: Hold one input to true)

2. Human memory and man-made memory have much in common

   (a) Compare human STM to dynamic random access memory (DRAM). How is DRAM refreshed? Can STM be refreshed? (HINT: DRAM is hardware, not biological; DRAM is dissipative; must be rewritten several times per second. DRAM words are shorter. STM is refreshed when it is admitted from a permanent long-term memory or is taken from the exactly same sensory image)
   (b) Compare human subconscious long-term memory to read-only memory (ROM). How is addressing accomplished in human memory; in ROM? (HINT: ROM is hardware, not biological. Certain types of ROM are easily erased, whereas human memory is long term. Human memory is associative; ROM generally requires addresses in the form of codes, for instance 16 bit addresses for 2^{16} words. These must be known beforehand. Human memory has many more words that can be located associatively)

3. Assume 32 memory cells with 32 binary attributes.

   (a) If each unique set of features represents exactly one memory recall, how many memories are held? (ANS. $2^{32} \approx 4$ billion)
   (b) If each unique set of features corresponds to two possible memories, how many memories are there? ($\approx$8 billion)
   (c) When multiple memories are activated, only one is returned. How would you identify those that are blocked out? (Those already returned are blocked out; those farthest away electrically tend to be blocked out until they are next in line to be returned)
   (d) How would you explain what happens during conflicting memory recalls, using the model of this chapter. (Multi-match circuitry takes over)

4. Illustrate a traditional associative memory with five words. Assume a database of automobile specifications that are retrieved according to make of auto, for example, Chevy, Ford, Dodge, Chrysler, and Cadillac entered randomly.

   (a) Compare an associative search to other kinds of searches. [HINT: try comparing to (1) random search and (2) sequential search from the top]

5. Redesign the memory subcircuit given in the chapter so that

$$\text{Rout} = \text{Rin}\ x' + \text{Rin}\ Q$$

   (HINT: Develop a circuit from this Boolean equation)

6. It was suggested that random search can result in cues for recall. Assume that you meet an old friend but cannot remember his name. However, your friends have the names Albert, John, Robert, Paul, Peter, Tom, and Zeke. Your old friend's name is Zeke!

   (a) To aid memory, begin searching names in alphabetical order from the top. How many trials might be required? (ANS. 7)

(b) Randomize the names and then search from the top, is recall faster or slower? How many trials are required for your new list? (HINT: Generally less than 7)

7. Provide an example of recall that is modeled after a classic car, in which you remember the chrome hood ornament and the red and white two-tone body but not the other features of the car. (HINT: Trace the recall process and signals in a four-cell model as done in the chapter's example)

Multiwrite Circuits

8. How is memorization enabled using the model of this chapter? (HINT: By rehearsal of an image that is not yet memorized)
9. During memorization how are multiple copies of the same image avoided? (HINT: a lockout signal is generated by the first return)
10. Devise a way to obtain a *Hit* signal from memory. (HINT: Refer to the circuits given in this chapter, especially Lout and Rout)
11. Devise a way to obtain a *NoHit* signal from memory. (HINT: A circuit for this has not been given, so it is up to the student to create one; perhaps use some sort of timer that is stopped by a hit, but which otherwise sends out a *NoHit* signal after a certain delay)

Simulated Qubits in Savant Memorization

12. Describe briefly the difference between memorization and learning as portrayed in this chapter. (HINT: Memorization is instant, learning requires time for synaptic development)
13. Provide a brief explanation of how a gifted savant can quickly memorize an entire book (HINT: Rehearsal filters enable state machine learning using fewer rehearsals)
14. Sketch a plan for a rehearsal filter that detects when steps in a procedure are repeated three times. (HINT: Use three toggle units in a cascade)

Chapter 9
Neuroquantology, the Ultimate Quest

Introduction

A circuits and systems view of consciousness is in a sense novel given the pervasive popularity of such sciences as molecular biology and biochemistry. But the fact remains that neurons are fundamentally a circuit, constrained to be connected in a certain way to work as a system. This system of course underlies human behavior.

The brain is not at all a general-purpose programmable computer, even though much effort goes into using them to simulate brains. Supercomputers, usually with a stored program, and often synchronized with the aid of a central clock, are funded and built a new one every year or so, as long as this author can remember, using microprocessors such as the currently popular ARM9.[1] Such machines are promoted as being electronic brains, and are always interesting, but thus far have failed to simulate consciousness, or even the beginnings of what practically any brain accomplishes with relatively little energy.

This book has introduced simulated qubits based on assemblies of recursive neurons while ignoring computational possibilities within individual neuron. The big question is: Is there more to brain power than pulsating neurons and synapses? In particular, do subneural particles have computational possibilities for supporting human intelligence and consciousness?

Possibilities are the following: (1) there might be a way for subneural molecules to configure themselves into simulated qubits to accomplish controlled toggling; (2) there might be a way for significant number of neural pulses to become synchronized together because of quantum mechanical tunneling, thus increasing

[1] ARM9, a 32-bit Reduced Instruction Set Computer (RISC) or Central Processing Unit (CPU), is useful for smaller conventional computers in mobile phones, robots, tablets, mobile phones, digital media and music players, handheld game consoles, calculators, and computer peripherals such as hard drives and routers.

J.R. Burger, *Brain Theory From A Circuits And Systems Perspective*, Springer Series in Cognitive and Neural Systems 6, DOI 10.1007/978-1-4614-6412-9_9,
© Springer Science+Business Media New York 2013

computational possibilities; and (3) there might be a way for molecules to support a quantum system, implying the possibility of teleportation and quantum algorithms within a brain. Such possibilities are not yet proven, but nevertheless they are too important to ignore.

Introduction to Neuroquantology

Currently there is considerable research and speculation about the possibility of quantum effects within a brain, with profound implications to intelligence and consciousness. This has given rise to the field of *neuroquantology*, a field that strives to reconcile neuroscience and psychological phenomena with the mysteries of quantum mechanics.

Quantum computations within neurons may help explain such human traits as common sense, truth judgment, intuition, artistic appraisal, and other hallmarks of human intelligence [1]. These, it is thought, would require orders of magnitude more computing power than what is possible using the most powerful computers known today.

By and large, neuroquantology is admittedly prescientific, but nevertheless it is a worthwhile field that benefits the scientific community. Quantum mechanics is a valuable metaphor for a variety of behaviors, and if nothing else, provides a useful vocabulary. It also has valuable proposals for future research. Certainly one can say that thinking about quantum mechanics exercises the brain, which is a healthy activity for those who think about it [2].

Quantum computing was suggested in the 1980s, suggesting that states of 0 and 1 (false and true) could be held probabilistically in a superposition within qubits, as introduced previously. Such qubits may be altered with reversible transforms, resulting in controlled toggling and other computationally useful events. Significantly, they are easily entangled and, upon measurement, are subject to novel nonlocal effects. But ultimately, when observed, each qubit is a simple 0 or a simple 1 as output.

By using appropriate transforms, difficult large-scale calculations become theoretically possible, although completely impossible classically. Quantum computations have immense potential for information processing, particularly since qubits can be small, possibly involving subatomic particles such as electrons. For instance 100 qubits may represent $2^{100} \approx 10^{30}$ numbers; these numbers are, in a sense, processed in parallel. Since qubits may have atomic dimensions, their promise is great; but they are not yet understood. Unsolved problems are (1) how to contain them in a coherent quantum system and (2) how to transform them in a practical way.

It is sometimes thought that small particles such as electrons are unimportant to biology, which deals mainly with larger molecules. For instance, texts on neuroscience hardly ever mention electrons. Nevertheless electrons are everywhere. They occur in molecules, atoms, and ions; they affect ion channels and synapses and other important biological interfaces. Electrons are inherently quantum mechanical. The question of the moment is: Does quantum mechanics affect biology in any direct way? Or is it just about larger classical molecules?

Tunneling

A clear separation between quantum physics and classical physics is in our understanding of where small particles are located. For example, in the everyday world an object is either inside a bottle or not. But for particles such as electrons that are very small (rest mass or weight is 9.11×10^{-31} kg), they may have a high probability of being inside their bottle, but a significant chance they are found outside their bottle. Quantum mechanically, a small object can appear outside its bottle without removing the top, breaking the bottle, and without punching a hole through it, or squeezing past its cork. This unfamiliar behavior is known as *tunneling*.

Tunneling, as predicted by quantum theory, is commonplace in the atomic world. Solid-state devices depend on it, for example, tunnel diodes. Alpha particles of radioactive decay are another common example; they suddenly pop out of a nucleus even though nuclear forces hold the nucleus together, and even though they do not have enough energy to get out. They tunnel out anyway. Small particles behave in a nonlocal manner, appearing where they are not supposed to be.

Copenhagen Interpretation

An accord reached in Copenhagen and promoted by Neils Bohr, a pioneer of quantum mechanics, is that humans know only what they can observe. This accord, the *Copenhagen interpretation*, is mainstream today and implies that physicists do not and cannot know the true nature of small particles. Quantum mechanics under this accord is a system of rules, no more, no less, amazingly successful rules based on observation, to predict what particles do.

In contrast, Heisenberg, another pioneer, proposed that waves of quantum probability really exist in nature, and that they evolve, adjust, and resolve (or collapse) to cause the events we observe in nature. This may be termed the *Heisenberg interpretation* under which probability waves as determined by Schrödinger's equation[2] are taken to be physical. Note, however, that such waves

---

[2] Schrödinger, a pioneer of quantum mechanics, contributed an equation which in one dimension is

$$i\eta \frac{\partial \psi}{\partial t} = -\frac{\eta^2}{2\mu}\frac{\partial^2 \psi}{\partial x^2} + V(x)\psi,$$

where $\psi(x, t)$ is a probability wave function that depends on time t; $V(x)$ is a potential field; and all else are given constants [11].

have not yet been observed directly. However, the probabilities they predict are accurate and supported experimentally.

The Heisenberg interpretation has been further developed [3] since it seemed that classical physics, taken from the bottom up, could not fully describe brain functioning. The nonlinearity of the synaptic system and the large number of states into which the brain can evolve, it was argued, point to a (top down) quantum mechanical system as the appropriate way to describe brain functioning. But if one proposes large-scale excitation of a brain in which conscious events are spontaneous selections (collapses) among possibilities (basic states), then fundamental questions arise. For instance, what computational power, exactly, creates the possibilities? And what are the possibilities? Such questions remain open.

Returning to tunneling, electrons in particular seem to have a role.

Electrons in Ionic Solutions

Electrons are easily captured by positive charge. Within ionic solutions they are routinely quasi-free but only for nanoseconds; because of their small mass, 1/1,836 of a proton, they achieve high velocities [4]; so they permeate a given region. Electrons tunnel through barriers where classically they cannot go, so in fact they permeate beyond any given region.

Electrons are plentiful in nature. The water molecule, for instance, has exposed eccentric electron orbitals from which electrons are easily freed. Stray electrons, knocked free by thermal agitation, are extremely mobile. Therefore it is reasonable to consider electrons when investigating some of the mysteries of neural electricity.

Clearly electrons have a major role in metals and semiconductors. At high electric fields, even insulators break down and conduct electrons. In contrast, neural membranes are surrounded by ionic solutions that mostly use ions to pass electrical charge, not electrons.

Even though ions pass nearly all of the electrical current for lower current densities, it is known that electrons in ionic solutions carry some of the current at higher current densities. The evidence is from the field of electroplating, where there is a decrease in electroplating efficiency as currents are increased. This is because some of the current is due to electron flow, which does not plate an electrode. Likewise in ion channels current densities are high, so ions may block the channel. Under such conditions electron tunneling may be important. Stray electron tunneling, aided by local electric fields, can be used to explain significant components of ion channel current.

Electrons in Ion Channels of Neurons

What makes the electron especially potent for its size is that it carries a unit of negative electrical charge ($-q = -1.6 \times 10^{-19}$ C, or coulomb). An interesting proposal is that electron tunneling into ion channels and subsequent capture by ions are important to the formation of the peak of a neural pulse [5].

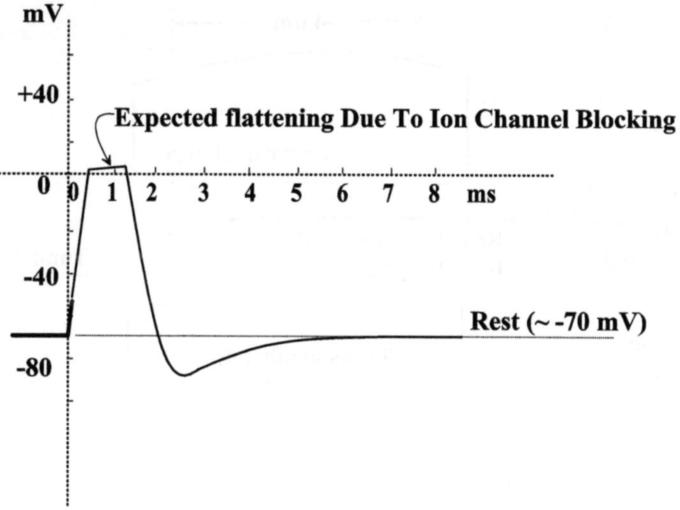

Fig. 9.1 Expected waveform distortion

When an ion channel opens, lower speed positive ions, which are by far the most plentiful, are stopped and reversed by a strong repelling electric field as soon as the internal neural voltage goes positive. A typical potential approaches +40 mV across the membrane; assuming a 4 nm long channel, this amounts to a repelling force of about +10MV/m, which is very high indeed, higher than the field for lightening in the sky (air breaks down at roughly 3 MV/m). Consequently ions are repelled, forced to drift back to where they came from. This should result in a sharp reduction, a distortion in the risetime of a neural pulse as internal voltage goes positive. Without electron capture the neural pulse might appear as in Fig. 9.1. But no such distortions are observed.

Pictured in Fig. 9.2 is an artist's rendition of an ion reversing direction in a channel, and capturing an electron as it leaves. Note that the electric field that repels outside positive ions also attracts inside electrons. Electrons still may diffuse into the channel if the electric field is not too high, or they may tunnel into the channel even for higher potential barriers. Captured electrons contribute to a neural pulse just like a transfer of positive ions, since losing negative charge is equivalent to gaining positive charge. Numerous such events would strengthen a neural pulse making it rise smoothly as commonly observed.

Once the possibility of tunneling is considered, this suggests other possibilities. For instance, depending on electric fields, stray electrons might tunnel through the phospholipid bilayer membrane without a transmembrane protein, resulting in charge transfer.[3] However, the possibility of direct tunneling through the bilayer membrane is not developed here.

[3] Probability of quantum penetration of stray electrons with energy 2 eV going through 1 nm of membrane barrier presenting a potential barrier of 5 eV is about 10^{-10}. Because electrons are so numerous, this is more than enough to build up significant amounts of charge.

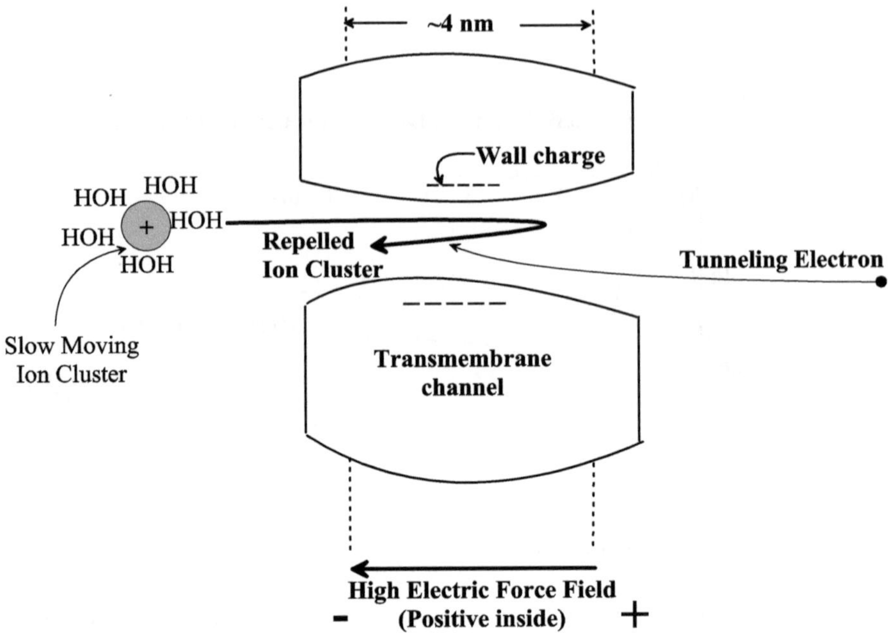

Fig. 9.2 Blocked ion channel

Electrons in Synapses

Postsynaptic retrograde electrons have been proposed as serving to release neurotransmitters in presynaptic vesicles [6]. Electrons are not usually mentioned in neuroscience. The usual simplified model for the release of neurotransmitters is that a positive pulse burst to the boutons in conjunction with thermal activity motivates vesicle release [7].

Walker proposed that an additional impetus is needed to open up the vesicles of neurotransmitters. Initially both pre- and postsynapse are at rest at about −70 mV. But when pulses arrive, each at +40 mV peak, positive charge tends to attract negative electrons across the synaptic cleft. The maximum energy possible for an electron is about $+40 - (-70) = 110$ mV. There are a great many stray electrons that could fly through the cleft. Thus a number of them hit the regions of vesicle gates, providing over a volt of energy (one electron-volt is 1.6×10^{-19} J, or joules). This is a big punch to small objects, proposed to be enough to nudge vesicle gates and to stimulate neurotransmitters.

Synapses are typically close together (refer to Fig. 9.3). The release of electrons from one may cause electrons to travel to others via quantum tunneling. Walker suggests that electrons tunnel to other synapses, instigating and effectively synchronizing synaptic firings. Tunneling is facilitated by a series of RNA (ribonucleic acid) molecules. This is very interesting, since the possibility of

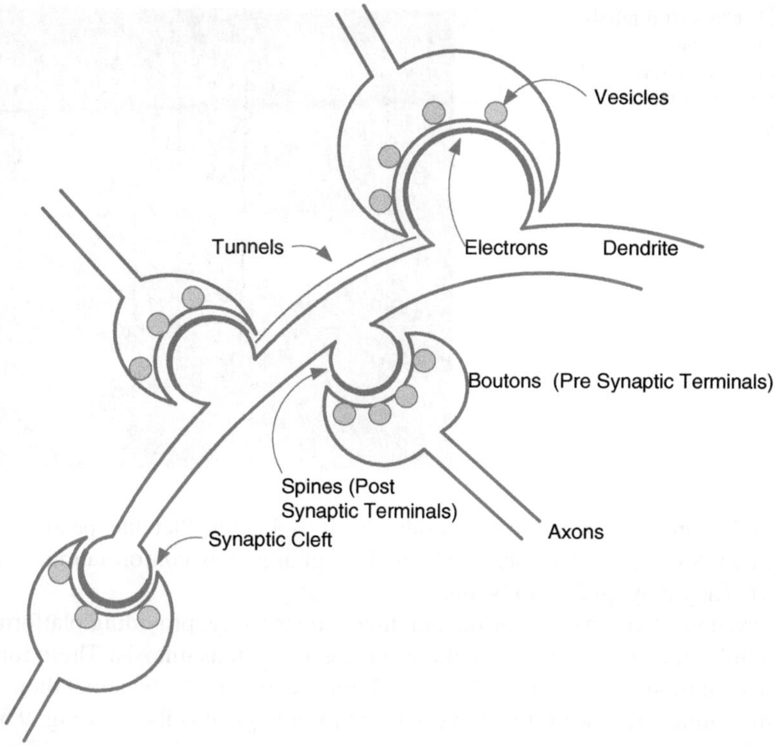

Fig. 9.3 Numerous synapses

synchronization would increase the power of neurons for high-level operations relating to consciousness. The possibility of synchronized synapses is important and cannot be ignored, but better proof is needed.

Opposition to Walker's proposal centers on the view that electrons are unnecessary to synaptic firing [8]. Thermal energy is claimed to be sufficient. Walker's proposal involves electrons that come from inside a neuron to control synaptic firing and neural pulse generation. What is inside has been proposed to contribute to human intelligence in other even more exotic ways.

Quantum Computations Inside Neurons and Microtubules

Penrose originally proposed a superposition of neurons "firing and not firing" that are capable of quantum computations [1]. Later Penrose and Hameroff proposed that microtubules (supporting structures in neurons) might hold qubits that interact and compute, and might even have a way to interact other neurons [9].

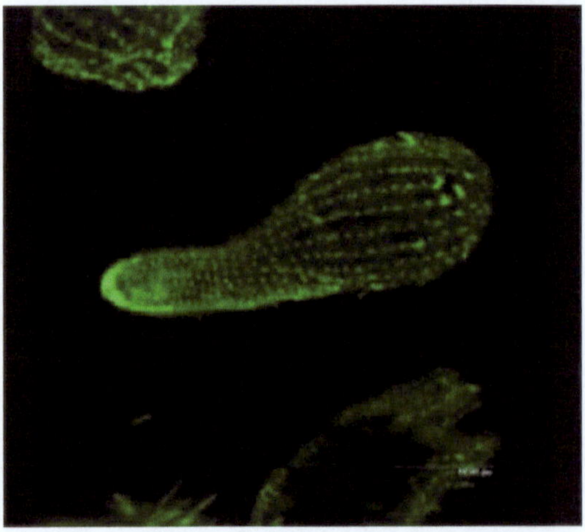

Fig. 9.4 Class III β-tubulin molecule (courtesy Wikipedia Commons http://en.wikipedia.org/wiki/Tubulin, accessed October 2012)

Tubulin proteins occur in microtubules. It is known that the peanut-shaped tubulin protein (Fig. 9.4) flexes about 30°, giving two conformational shapes (states). They may qualify to be qubits.

Microtubules are mainly for maintaining cell structure, providing platforms for intracellular transport and other cellular processes such as mitosis. These ropelike polymers of tubulin can grow as long as 25 μm and are highly dynamic; they, along with microfilaments, seem to be everywhere in biological cells as in Fig. 9.5. The outer diameter of a microtubule is about 25 nm, pictured in Fig. 9.6. There are roughly 10^7 tubulins per neuron.

It is proposed by Hameroff that quantum states in one neuron could interact with quantum states in other neurons via electrotonic gap junctions. These are conductive junctions through the glial between neurons. If quantum states can extend over many neurons, a large-scale system may be possible.

Partial justification for such a system is found in observations of gamma synchrony (a property of brain waves). These indicate that information may propagate through the brain much faster than ordinary neural bursts would permit. The implication is that entangled qubits are at work using quantum teleportation.[4]

The relevance of tubulin qubits is disputed on the grounds that quantum coherence is predicted to be less than a picosecond, that is, 10^{-12} s [2]. This is far too short to be of practical value in a neural environment. To refute this, Hameroff and Penrose recalculated that coherence may be for up to about a second, using a combination of electrostatic (Debye) screening, quantum error correction, and processes related to cellular maintenance [10].

[4] Two entangled qubits may have a state vector that looks like $|\psi>=\eta(|\mathbf{0}>|\mathbf{0}>+|\mathbf{1}>|\mathbf{1}>)'$. If one qubit is observed to be one, for example, then the other will be forced to be a one when it is observed, no matter where it is located. But if the first is observed to be zero, the other will be forced to be zero when it is observed. Note that a coherent quantum system is necessary. Teleportation is a nonlocal quantum property.

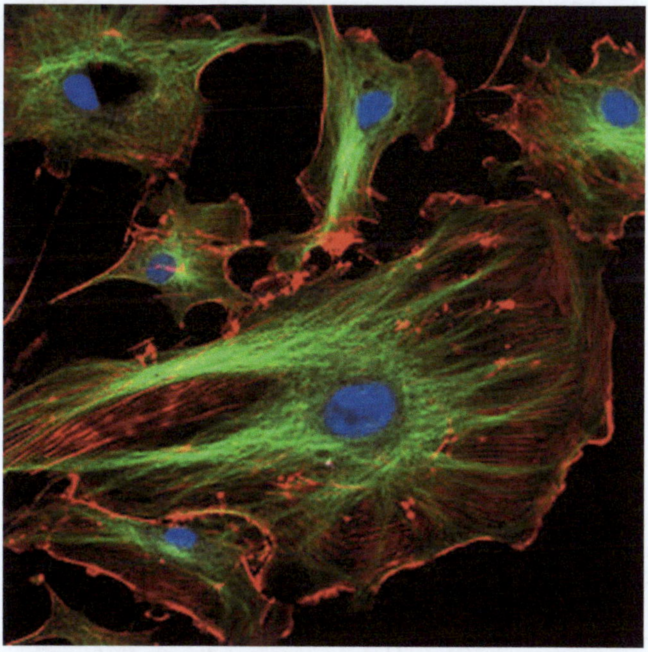

Fig. 9.5 Microtubules, filaments, and nuclei (courtesy Wikipedia Commons http://en.wikipedia.org/wiki/Cytoskeletal, accessed October 2012)

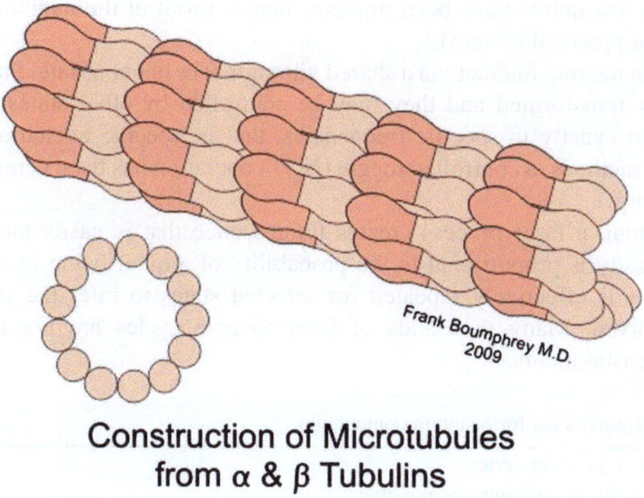

Fig. 9.6 Microtubule structure for a short segment (courtesy Wikipedia Commons http://en.wikipedia.org/wiki/Microtubule, accessed October 2012)

Hameroff and Penrose went on to develop an elaborate concept of consciousness. Underlying their concept is the Heisenberg proposal that probability waves have a reality of their own and are not just a way to explain the results of experimental observations. Hameroff and Penrose developed what they term *orchestrated objective reduction* (ORCH OR), a sort of spontaneous observation, after which the results of quantum-level calculations are manifest. A full explanation of ORCH OR may be found in the references.

Requirements for Quantum Computations

To have a quantum computer there are basic requirements [11, 12] as listed in Table 9.1.

Requirement 1

Quantum information must be represented in a quantum system: To represent quantum information, particles must be part of a quantum system with regular quantum states.

Trapped ions can constitute a quantum system, but only if reduced in temperature to minimize the interference of thermal vibration. They work with the lowest levels of phonons (vibrational states) so that it is possible to access the hyperfine spin states, that is, electron plus nuclear spin combinations. Quantum computers with only a few qubits have been implemented as proof of the quantum principle (but not as a practical system).

Ions in an ion trap interact via a shared vibrational or phonon state. States may be individually transformed and they may be controlled by other states using laser beams tuned exactly to specific frequencies, that is, specific excitation energies. Using such methods, a controlled toggle (NOT) operation has been demonstrated in the laboratory.

For readout, a laser process creates fluorescence that is easily analyzed, and whose intensity is proportional to the probability of a qubit being in a zero state. This process is effectively repeated for selected states to infer the states of the qubits involved. Many thousands of laser-observe cycles are possible, giving accurate measurements.

Table 9.1 Requirements for quantum computation

There must be a quantum system
Initialization of the system must be possible
Quantum information within the system must be appropriately transformable
Useful information must be available to the outside world

To mimic trapped ions, tubulins would require minimal thermal vibration, usually achieved for ions with liquid helium cooling, although there may be other ways. Alternately, there may be some other type of quantum system, as yet undiscovered, other than combined phonon-spin states.

Requirement 2

Initialization must be possible: In the case of trapped ions, they are cooled using processes termed *trapping* and *optical pumping* to put them into their thermal and spin ground states. Subsequently, states would have to be selectively excited with lasers to represent data. This cannot be done in practice yet for more than a few qubits.

Initialization is a fundamental issue; a system will not work without proper initialization. In the case of tubulins, it might be reasonable to assume that they naturally decay to a zero state. Then each unit merely needs to be excited to be a true for those true attributes, to represent significant amounts of data.

Attributes of an image might be entered at the synapses of a given neuron. Then each synapse might exert control so as to set, as required, each corresponding internal tubulin to true.

If corresponding synapses and tubulins are fairly close physically, special-purpose receptors could send signals, or possibly electrons to tubulins. It is at least conceivable that tubulins can be initialized to hold a mental image, although, of course, this has never been verified experimentally.

Requirement 3

Quantum information must be appropriately transformable: Ions in ion traps are manipulated using exactly positioned and finely tuned laser beams, giving, we are told, most of the standard qubit operations. Practical calculations with qubits would necessitate a long sequence of laser operations. For tubulins, smaller traveling molecules with special codes have been proposed and this concept awaits experimental verification.

The amount of required qubit manipulation is significant. When a previous chapter studied simulated qubits, a sequence of codes had to be taken from a special read-only section of long-term memory. If images can be initialized onto microtubules, then perhaps manipulation codes, taken from long-term memory, can also be applied. This has not been observed experimentally, but it is a possibility.

Requirement 4

The readout method must be appropriate: Readout in a quantum computer under the Copenhagen interpretation means the qubits must be physically observed. At the time of observation they each will "collapse" into $|0>$ or $|1>$ according to their probabilities.

Output is often held by a small number of qubits, for example, three or four to represent image priority. These qubits presumably will interact with regular neurons, causing them to produce action potentials.

Observation of the states of trapped ions requires a detailed observation of the system by judicious laser probing. This would not work within a human brain, but perhaps there is some unknown not yet identified way to make an observation.

Self-collapse has been considered. This implies that conscious short-term memory will have forced upon it a new image at the convenience of mysterious internal forces.

Simulated Qubit Proposal Using Tubulins

Perhaps more likely than the qubits of quantum physics are simulated qubits composed of tubulins. They might communicate with each other via tunneling electrons between synapses, resulting in a controlled toggle. For this to work, one or more controlling (*fm*) synapses must detect their tubulin qubits to be true; and then synapses must send a signal to objective (*to*) toggles to change or flip their states. Operations in this scenario would be orchestrated by code obtained from specialized long-term memory source, all subliminal.

Conclusions

This chapter introduces the possibility that subneural particles and electrons play an important role in neural systems. Electrons are proposed to be important within ion channels, especially once internal voltage goes positive and repels outside ions. Channels are blocked to most ions, but stray electrons may readily tunnel into a channel and recombine to transfer charge. This permits voltage to reach the peak of a neural pulse.

Introduced in this chapter is a proposal that electrons move from post- to presynaptic regions, providing significant energy essential to release neurotransmitters from presynaptic vesicles. Such electrons, which may be quite numerous, are proposed to tunnel and trigger nearby synapses within their range. Thus they may play an important role in synchronizing neural operations leading to consciousness.

A proposal is introduced in this chapter under which tubulin molecules form qubits for quantum computations within a neuron. A large number of such qubits may be possible with a significant impact on the computational abilities within a brain.

On the other hand, simulated qubits at the molecular level would not require a quantum system, and would be less prone to decoherence. This leads to the possibility that subneural tubulin proteins might simulate qubits and thus might constitute powerful computations for the purpose of explaining the more advanced of human abilities.

It should be noted that the systems mentioned in this chapter are mainly prescientific, meaning that additional experimentation is needed to prove or disprove them.

References

1. Penrose R (1989) The emperor's new mind: concerning computers, minds and the laws of physics. Oxford University Press, Oxford
2. Tegmark M (2000) The importance of quantum decoherence in brain process. Phys Rev E 61:4194
3. Stapp HP (1993) Mind, matter and quantum mechanics. Springer, Berlin
4. Conway BE (1981) Ionic hydration in chemistry and biophysics. Elsevier Scientific Publishing Company, Amsterdam
5. Burger JR (2010) The electron capture hypothesis—a challenge to neuroscientists arXiv:1006.1327, Sept 2010
6. Walker EH (2000) The physics of consciousness: the quantum mind and the meaning of life. Perseus, Cambridge, MA
7. Cowan WM, Sudhof TC, Stevens CF (2001) Synapses. John Hopkins, Baltimore, MD
8. Donald MJ (2001) Book review, the physics of consciousness: The quantum mind and the meaning of life. Psyche, vol. 7, www.bss.phy.cam.ac.uk/~mjd1014, accessed March 4, 2013
9. Hameroff S, Penrose R (2003) Conscious events as orchestrated space-time selections. NeuroQuantology 1:10–35
10. Hameroff S (2007) Orchestrated reduction of quantum coherence in brain microtubules: a model for consciousness. NeuroQuantology 5:1–8
11. Merzbacher E (1961) Quantum mechanics. Wiley, New York, p 35
12. Nielson MA, Chuang IL (2000) Quantum computation and quantum information, Cambridge series on information and the natural sciences. Cambridge University Press, Cambridge [Paperback]

Self-Study Exercises

(If you desire to keep a record of your efforts, please type your answers.)

1. Contrast Bohr's Copenhagen interpretation to the Heisenberg speculation.
2. Stapp was perhaps the first to suggest a quantum theory of consciousness. State a strength and a shortcoming in Strapp's theory.
3. Describe Walker's theory of consciousness. State a strength and a shortcoming in Walker's theory.
4. Describe the Hameroff–Penrose theory of consciousness. State a strength and a shortcoming in their theory.
5. Explain (and discuss briefly) the four requirements for a quantum computer.
6. Explain quantum teleportation in technical terms.

Chapter 10
The Phase of the "1"

Introduction

Electrical variables can carry a great deal of information, including waveform shapes, durations, amplitudes, and phases. Previous chapters have mostly ignored the phase variable for the waveform generated by a single recursive neuron. This chapter introduces an auxiliary recursive neuron to operate at frequency f_1, which is the frequency for a logic 1. The purpose of the auxiliary recursive neuron operating at f_1 is to register a relative delay termed *Delay* β, measured in microseconds, which gives a phase angle β measured in degrees [1].

Delay β comes into play mainly when simulated qubits have been initialized to have equal probabilities for false and true, although other situations are possible. Applications considered below are as follows: (1) β works to identify an unknown binary function of a single input, as held by a simulated qubit for which direct observation is impractical. (2) β works to identify an unknown binary function of many inputs, as held by many simulated qubits for which direct observation is impractical. Other applications mentioned are as follows: (3) this chapter discusses a satisfiability problem in which it is desired to identify inputs that satisfy a given (known) Boolean multivibrator function of many simulated qubits, using a minimum number of observations.

Finally, (4) this chapter investigates a way to use recursive neurons to increase the density of stored information in long-term memory. This may be done by packing data into a state vector based on simulated qubits, each of which is either false with 100 % probability, true with 100 % probability, or both true and false with equal 50 % probabilities.

Biological applications, if any, are not the focus of this chapter. Mainly this chapter aids in understanding certain algorithms for simulated qubits, and develops a pantheon of noteworthy operations that simulated qubits might accomplish.

J.R. Burger, *Brain Theory From A Circuits And Systems Perspective*, Springer Series in Cognitive and Neural Systems 6, DOI 10.1007/978-1-4614-6412-9_10,
© Springer Science+Business Media New York 2013

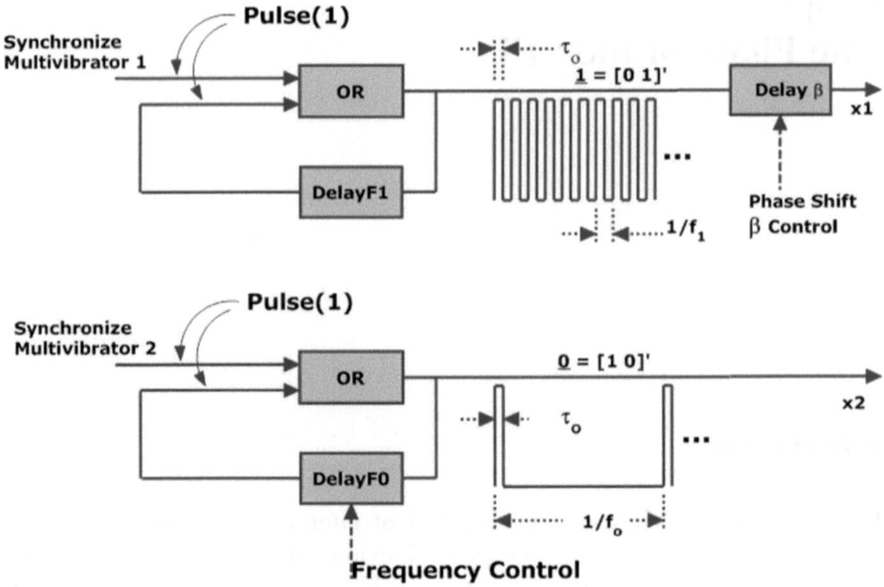

Fig. 10.1 Multivibrators representing true (top) and false (bottom)

The Phase of the 1

Figure 10.1, top part, suggests an extra multivibrator that is dedicated to representing phase at frequency f_1, the 1 state. The phase of the 1 at frequency f_1 is adjusted via *Delay* β in the figure, and is relative to *Delay* $\beta = 0$. So this phase is also relative to the first pulse of other multivibrators, whose phase shift is zero. Multivibrator waveforms in this chapter are assumed to be harmonically related in an attempt to increase the tractability of the analysis. For example, if the low frequency is 1 Hz, then frequency can step up to its maximum, perhaps 400 in 1 Hz steps. Multivibrators are synchronized by triggering them at the same instant.

Note that the waveform for f_1 in the figure is shown as being 50 % duty cycle, mainly because 50 % is easy to illustrate. It could happen that duty cycle for the f_1 waveform differs. This is not important below because the only purpose of the f_1 waveform is to register phase.

To simulate a qubit more accurately, the 0 and the 1 must be combined, or held in superposition. If the media were linear, signals might simply be added producing two tones at once, as is common mechanically. Unfortunately, neurons are far from linear, so linear combinations will not work. But two signals may be combined logically, for instance, in an AND gate as in Fig. 10.2, although other logical combinations may also serve. Note that x_1, x_2 are waveforms from the above figure.

Assuming about equal probabilities for false and true, a composite waveform (not drawn to scale) is portrayed in Fig. 10.3. This waveform results by increasing

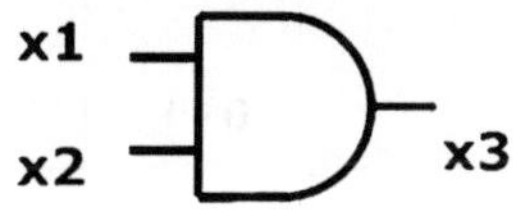

Fig. 10.2 AND gate

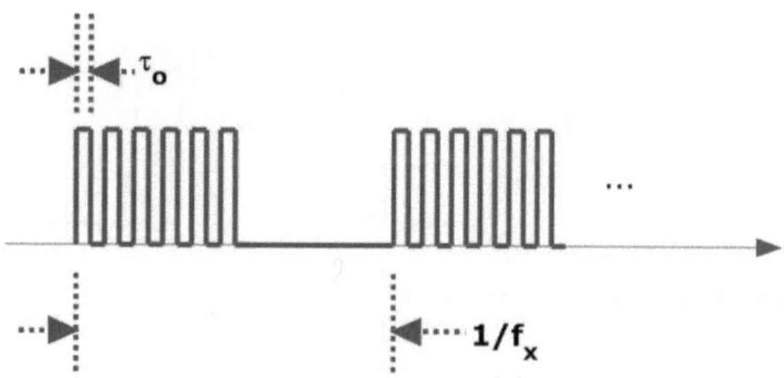

Fig. 10.3 Combined waveforms with no phase shift

the frequency of multivibrator 2 to some f_x while maintaining the frequency f_1. Effectively, the waveform for f_1 is modulated by the waveform for f_x.

To summarize, the probability of a true is determined by f_x as in previous chapters. The phase factor, used below, is held by f_1. The availability of relative phase, the phase of the 1 relative to the phase of the 0, identified above as phase β, creates interesting possibilities for multivibrators.

Waveforms for [1 1]′ and [1 −1]′

A simulated qubit can be expressed as a combination of true and false by initializing a multivibrator $\underline{\mathbf{a}}$ as follows: $\underline{\boldsymbol{a}} = \eta\ [1\ 1]'$. Although not needed at this time, it is educational to note that this vector, like any linear vector, can be analyzed as $\underline{\boldsymbol{a}} = \eta\ (\underline{\mathbf{a_1}} + \underline{\mathbf{a_2}})$:

$$\underline{\mathbf{a_1}} = \underline{\mathbf{0}} = \begin{bmatrix} 1 \\ 0 \end{bmatrix}, \quad \underline{\mathbf{a_2}} = \underline{\mathbf{1}} = \begin{bmatrix} 0 \\ 1 \end{bmatrix}. \tag{10.1}$$

To increase understanding, $\underline{\boldsymbol{a}} = \eta\ [1\ -1]'$ may be expressed as

$$\underline{\mathbf{a_1}} = \underline{\mathbf{0}} = \begin{bmatrix} 1 \\ 0 \end{bmatrix}, \quad \underline{\mathbf{a_2}} = -\underline{\mathbf{1}} = \begin{bmatrix} 0 \\ -1 \end{bmatrix}. \tag{10.2}$$

The $\underline{\mathbf{0}}$ and $\underline{\mathbf{1}}$ are termed *basis vectors*. For instructional purposes, waveforms for [1 1]′ and [1 −1]′ are suggested in Fig. 10.4.

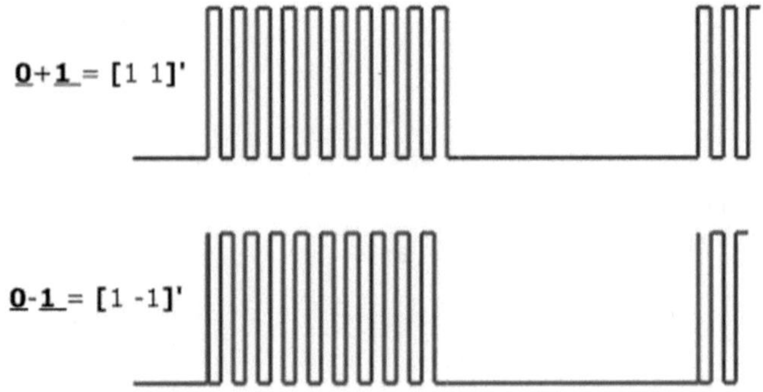

Fig. 10.4 Phase shift in the "1"

The h-Transform

The problem addressed now is to distinguish a simulated qubit in state η [1 1]′ from one in state η [1 −1]′. It is not permitted to place an oscilloscope probe directly to a multivibrator because recursive neurons are assumed to be too small, delicate, and easily upset. Rather it is desired to perform some sort of transformation on the qubit such that after sampling occurs, phase will be indicated as an output of the sampling.

By modifying delays, it is possible to manipulate the frequency and the phase of the multivibrators. This in turn can be made to provide useful information after sampling, which results in a simple true or false answer about the hidden states of the multivibrator vectors. Manipulating frequency and phase is an operation on the probability space of a simulated qubit, that is, a rotation on the sphere of a previous chapter. An example is given by the h-transform, which is analogous to the Hadamard transform of mathematics.

Let $\underline{\mathbf{a}} = \eta\ [1\ 1]'$ be a point on the sphere. Based on this, define $[p\ q]'$ where $p = 1; q = 1$. Next consider a transformation that provides $\underline{\mathbf{b}} = [b_1\ b_2]'$:

$$b_1 = (1 - \eta^2)p + \eta^2 q = 1. \tag{10.3}$$

$$b_2 = (1 - \eta^2)p - \eta^2 q = 1 - 2\eta^2 = 1 - 1 = 0. \tag{10.4}$$

Note that $\eta^2 = 0.5$. So $\eta\ [1\ 1]'$ transforms to $[1\ 0]'$ internally, which after sampling is $\underline{\mathbf{0}}$ or false, determined with certainty with one sampling (or measurement). So it is possible to know that the original vector is $\eta\ [1\ 1]'$ without having to probe and perhaps disturb the multivibrator. Finally the h-transform is reversible in case it is desired to return the vector to its original state.

For $\eta\ [1\ -1]'$, similarly define $[p\ q]'$ where $p = 1; q = -1$. Then consider a transformation that provides $\underline{\mathbf{b}} = [b_1\ b_2]'$:

Table 10.1 The four functions of a single binary variable

x_i	g_0	g_1	g_2	g_3
0	0	0	1	1
1	0	1	0	1

Table 10.2 Procedure to identify constant and nonconstant functions

x_i	g_2	Prepared list	Tagged list	Assumed list structure
0	1	1	−1	$=a_1$
1	0	1	1	$=a_2$

$$b_1 = (1 - \eta^2)p + (\eta^2)(q) = 1 - 2\eta^2 = 0. \tag{10.5}$$

$$b_2 = (1 - \eta^2)p - (\eta^2)(q) = 1. \tag{10.6}$$

Since $\eta^2 = 0.5, \eta\,[1\,-1]'$ transforms to $\underline{\mathbf{1}} = [0\ 1]'$ which after sampling is 1 or true, determined with certainty with one sampling (or measurement). So a system can know that the original vector is $\eta\ (1\ -1)'$ without having to probe and perhaps disturb the multivibrator. This transform is reversible. For example, $[0\ 1]'$ may be transformed to be $[p\ q]'$ where $p = 0$; $q = -1/\eta$, returning the vector to $\underline{\mathbf{b}} = \eta\ [1\ -1]'$.

Binary Function Classification Involving One Simulated Qubit

Assuming an ordinary binary function of a single bit, possibilities are, the function is constant: either $g(x_i) = 0$ or $g(x_i) = 1$ for $i = 1, 2$; (note that $x_1 = 0, x_2 = 1$) otherwise it is nonconstant: either $g(x_i) = x_i$ or $g(x_i) = x_i'$, the complement of x_i. This information is presented in a truth table, Table 10.1.

Two functions, g_0 and g_3, are constant functions as x_i varies; the other two, g_1 and g_2, are nonconstant.

Recall that simulated qubits have at their core a set of multivibrator waveforms operating at a certain frequency and phase. A *multivibrator function* is a sequence of rotations, polar and azimuthally, of the probability vector that introduces negative signs into the presampled output possibilities. For example, two simulated qubits $[a_1\ a_2]'$, $[b_1\ b_2]'$ may be prepared to be $[1\ 1]'$, $[1\ 1]'$ meaning that each has 50 % probability of being true and 50 % probability of being false. Overall there are finite probabilities for the following combinations a_1b_1, a_1b_2, a_2b_1, a_2b_2. A multivibrator function is defined to be one that places a negative sign on one more of these terms.

Generalizing, for any number of simulated qubits, the locations of the negative signs are congruent to the 1s of a particular Boolean function. Negative signs are invisible in the readout method, so with simple sampling fails to identify the multivibrator function should it be unknown. A special procedure is required to identify or at least classify the function. For example, refer to the inverter function, g_2, in Table 10.2.

Imagine a qubit that is prepared to be $\eta\,[a_1, a_2]' = \eta\,[1, 1]'$; this is represented by a *Prepared List* in the table; in this case, $a_1 = 1, a_2 = 1$. A function is applied in such a way so as to tag the prepared list with negative signs at those inputs for which the function is true. That is, the 1s in the prepared list are converted to -1s in those rows where the truth table for g_2 has a 1. In this case the top element is tagged (refer to the *Tagged List*).[1]

Processing is now applied to see the effects of the function on the original prepared qubit which had $a_1 = 1, a_2 = 1$. Variables in the *Assumed List Structure* are equated to the *Tagged List* and solved to determine that $a_1 = -1, a_2 = 1$, the solution of which is trivial for a single qubit but more involved for many qubits. The qubit is assumed to be transformed to $a_1 = -1, a_2 = 1$. The overall result of the above multivibrator function is to change the phase of the simulated qubit from $\eta\,[1, 1]'$ to $\eta\,[-1, 1]'$. This phase change does not affect the probabilities of what is sampled and read out; true will occur 50 % of the time, and false 50 % of the time.

It is desired to use an h-transformation. Upon h-transformation, the result $-\underline{\mathbf{1}} = [0\ -1]'$ occurs in the multivibrator. When this is sampled it is observed to be a true. Note that the minus sign in $[0\ -1]'$ does not affect the truth value. Either $+\mathbf{1}$ or $-\mathbf{1}$ internally will give a true after sampling.

Using the above procedure, it is soon discovered that any *nonconstant function* will result in the observation of a true, and that any *constant function* will result in the observation of a false.

The above demonstration, although elementary, leads to an interesting, if subtle, consequence for simulated qubits that hold false and true simultaneously. If an unknown multivibrator function has been applied to a simulated qubit, there is an easy way to identify that unknown binary function. As explained above, it is not permitted to place an oscilloscope probe directly the multivibrator in some sort of direct observation.

To help identify the function, probability processing as above is used. After applying the h-transformation, the multivibrator vector may end up being $\underline{\mathbf{a}} = \pm\,[1\ 0]'$. This means that the sampling readout will be false with certainty, so it may be concluded that the multivibrator function is constant, either $g(a_i) = 0$ or $g(a_i) = 1$ for $i = 1, 2\ (a_1 = 0, a_2 = 1)$. But if after h-transformation the result is $\underline{\mathbf{a}} = \pm\,[0\ 1]'$ then the sampled readout is true with certainty, and it may be concluded that the function is nonconstant, either $g(a_i) = a_i$ or $g(a_i) = a_i'$.

Constancy or non-constancy is a *global* property, which can be determined, as above, by probability transformation and then observing $\underline{\mathbf{a}}$ to be $\underline{\mathbf{0}}$ or $\underline{\mathbf{1}}$. Normally, function classification for constancy or non-constancy usually requires evaluations of the function at least two times, once for an argument of 0 and once for an argument of 1. But using multivibrator functions, only one observation is necessary to make this determination.

[1] Note that a function is implemented by operations that change the sign of selected terms in the Prepared List to become the Tagged List.

The above method is a variation on *Deutsch's quantum algorithm* [2]. For larger numbers of simulated qubits additional global properties of multivibrator functions become identifiable.

Symmetric and Antisymmetric Multivibrator Function Classification Using *n* Simulated Qubits

Given n synchronized multivibrators, a multivibrator function operates on each possible combination for which there is a probability. For example, for $n = 2$, there could be a finite probability for each combination a_1b_1, a_1b_2, a_2b_1, $a_2\ b_2$.

The class of functions g_{sa} (symmetric and antisymmetric functions) can be identified by their patterns in truth tables [3]. Symmetric means that the entries below the center of the truth table is a mirror image about the center of the half above the center. Antisymmetric means that a bitwise complement of the second half is a mirror image about the center of the first half.

A symmetric or antisymmetric function g_{sa} of n binary variables with dimension $k = 2^n$ has the property of being symmetric or antisymmetric about the center of its truth table and being symmetric or antisymmetric in each binary subdivision of 2^{n-i} entries of the table $i = 1, 2, \cdots, n - 1$ for $n > 1$. This class of binary functions has an even number of true entries in its truth table for $n > 1$; $n = 1$ is a limiting case for constant and non-constant binary functions of one bit as above; any binary function with $n = 1$ is either symmetric or antisymmetric. Examples for $n = 2$ are provided in Tables 10.3 and 10.4.

Assume that a multivibrator function in this class has been applied to the probability space of n multivibrators but that the actual function is unknown. As above, it is not permitted to place an oscilloscope probe directly into a recursive neuron because it is assumed small, delicate, and easily upset. A procedure for

Table 10.3 $g_{sa} = x_1 \oplus x_2$, symmetric

Truth table		
x_1	x_2	g_{as}
0	0	0
0	1	1
1	0	1
1	1	0

Table 10.4 $g_{sa} = x_1$, antisymmetric

Truth table		
x_1	x_2	g_{as}
0	0	0
0	1	0
1	0	1
1	1	1

Table 10.5 Applying $g_{sa} = x_1 \oplus x_2$ to multivibrator states

Truth table					
x_1	x_2	g_{as}	Prepared list	Tagged list	Assumed list structure
0	0	0	$a_1b_1 = 1$	1	$=a_1b_1$
0	1	1	$a_1b_2 = 1$	−1	$=a_1b_2$
1	0	1	$a_2b_1 = 1$	−1	$=a_2b_1$
1	1	0	$a_2b_2 = 1$	1	$=a_2b_2$

identifying an unknown g_{sa} is illustrated for the case $n = 2$ in Table 10.5. The two multivibrators are represented by $[a_1\ a_2]'$, $[b_1\ b_2]'$. The two multivibrators are prepared to have equal probability and zero phase as suggested by the *Prepared List* in the table. This list is ordered like a binary count merely for convenience: $a_1b_1, a_1b_2, a_2b_1, a_2b_2$. This form is inspired by the terms of a direct product $\underline{\mathbf{a}} \otimes \underline{\mathbf{b}}$. Probability normalization factors have been ignored in the tables to simplify the notation.

The end result of applying the multivibrator function g_{sa} is selected phase reversals in $a_1b_1, a_1b_2, a_2b_1, a_2\,b_2$, as suggested by the *Tagged List*. The locations of the negative signs correspond to the 1s in the truth table under g_{as}. This is what the function does; it selectively tags the combinations $a_1b_1, a_1b_2, a_2b_1, a_2\,b_2$ with negative signs. This in turn results in certain sign changes within the qubits $[a_1\ a_2]'$, $[b_1\ b_2]'$. A solving procedure is a simple way to determine what these sign changes are.

First equate the *Assumed List* entries to the *Tagged List* entries, for example $a_1b_1 = 1$, $a_1b_2 = -1$, $a_2b_1 = -1$, $a_2b_2 = 1$. Begin by assuming that $a_1 = 1$. Solve the equations to obtain $a_1 = b_1 = 1$, $a_2 = b_2 = -1$. Then $[a_1\ a_2]'$, $[b_1\ b_2]' = [1\ -1]'$, $[1\ -1]'$; this in turn can be transformed to $[0\ 1]'$, $[0\ 1]'$ using h-transforms, which upon sampling can be observed as $\underline{\mathbf{1\ 1}}$, and will be $\underline{\mathbf{1\ 1}}$ with certainty each time it is observed. This classifies the function, since the math is such that only symmetric and antisymmetric functions with truth tables "0110" or "1001" would be observed to be $\underline{\mathbf{1\ 1}}$.

A function with the truth table "1 0 0 1" is also symmetric (about its center). Prepared multivibrators can be transformed as in the above table to become $[-1\ 1]'$, $[1\ -1]'$. This, in turn, transforms to $-[0\ 1]'$, $[0\ 1]'$ but this is still observed to be $\underline{\mathbf{1\ 1}}$. A negative sign associated with $\underline{\mathbf{1\ 1}}$ would not be seen by sampling, because sampling does not respond to phase.

Functions such as "0 1 1 0" and "1 0 0 1" are termed *complementary* functions. Complementary multivibrator functions will give the same sampled output. Thus a sampled result of $\underline{\mathbf{1\ 1}}$ classifies a given function to be "0 1 1 0" or its complement "1 0 0 1" within the class of symmetric and antisymmetric functions.

As another example of this, consider Table 10.6, which is a function with a truth table "0 0 1 1" and so is antisymmetric; it transforms $[1\ 1]'$, $[1\ 1]'$ into $[1\ -1]'$, $[1\ 1]'$. Then it transforms to $\underline{\mathbf{1\ 0}}$. This ultimately classifies the function as being either "0 0 1 1" or "1 1 0 0."

Table 10.6 Applying $g_{sa} = a_2$ to multivibrator states

Truth table					
x_1	x_2	g_{as}	Prepared list	Tagged list	Assumed list structure
0	0	0	1	1	$=a_1b_1$
0	1	0	1	1	$=a_1b_2$
1	0	1	1	−1	$=a_2b_1$
1	1	1	1	−1	$=a_2b_2$

Table 10.7 Decoding function $g_d = x_1'x_2$

Truth table					
x_1	x_2	g_{as}	Prepared list	Tagged list	Assumed list a_ib_j structure
0	0	0	1	1	a_1b_1
0	1	1	1	−1	a_1b_2
1	0	0	1	1	a_2b_1
1	1	0	1	1	a_2b_2

For n multivibrators, there are 2^n symmetric and antisymmetric multivibrator functions whose truth table begins with a 0, and 2^n symmetric and antisymmetric "complementary" multivibrator functions whose truth table begins with a 1. Such multivibrator functions can be classified with certainty to within a complement, using only one evaluation. Normally a function of n binary variables would require up to 2^n calls to the function to fill in its truth table. So if an unknown multivibrator function within the class g_{sa} is applied to simulated qubits, it can be classified to within a complement with only one call to the function and only one observation.

The Satisfiability Problem for a Multivibrator Decoding Function

An interesting class of binary functions g_d may be defined on integers k, $0 \leq k \leq 2^n - 1$, so that $g_d(k) = 0$ for all k except for $k = k_o$, for which $g_d(k_o) = 1$. This defines what are known in electrical engineering as a *decoding function*. Assume that g_d is known and given but that what satisfies it, k_o is unknown. If one searches classically, evaluating $g_d(k)$ until one finds k_o one might have to make as many as 2^n evaluations of $g_d(k)$.

When a binary function is known, it is not always easy to discover what satisfies it, or makes it true, especially for complicated or indirectly presented functions. Finding what satisfies a given binary function such as $f_d(k)$ is known as a *satisfiability problem*.

Satisfiability problems also occur for multivibrator functions. Here is a simple simulated qubit example in which the normalization factors have been omitted in order to focus on the logic. Consider a multivibrator function on a Prepared List that provides the Tagged List in Table 10.7. Note that the truth table is true only for an input of 01, so this is a decoding function; it satisfies the condition that only one

Table 10.8 Analysis for $g_d = x_1{}'x_2$

Tagged list	Assumed list structure	Mod1 permutated tagged list	Mod2 1st entry sign reversal
1	a_1b_1	1	−1
−1	a_1b_2	1	1
1	a_2b_1	−1	−1
1	a_2b_2	1	1

code satisfies the function g_{as}, and that there is only one negatively tagged entry in the Tagged List. Now, to demonstrate the method, it must be assumed that what satisfies the multivibrator function is unknown.

Simulated qubits, n in number, can be prepared to provide all 2^n combinations of true and false. It is assumed that a multivibrator function will tag a given combination with a minus sign where the function is satisfied, as suggested in the table. The location of the tag cannot be determined by direct sampling because negative signs do not affect the probability of a given output. Electrical probing of the waveforms is assumed to be impossible.

In an attempt to discover which a_ib_j term has the negative tag, a transformation is going to be applied prior to sampling and readout. The transformation is assumed available for a multivibrator function. First the Tagged List is permutated using an **H** matrix of **H** matrices, or $\mathbf{H}^{\otimes 2}$ giving the result *Mod1* shown in Table 10.8. Note that

$$\mathbf{H} = \frac{1}{\sqrt{2}}\begin{bmatrix} 1 & 1 \\ 1 & -1 \end{bmatrix};\ \mathbf{H}^{\otimes 2} = \frac{1}{\sqrt{2}}\begin{bmatrix} \mathbf{H} & \mathbf{H} \\ \mathbf{H} & -\mathbf{H} \end{bmatrix}.$$

This list is modified again with a transformation that reverses only the phase of the first entry a_1b_1 as shown in the right of the table under *Mod2*.

There are now an even number of negative signs in the list under Mod2 in the table. Having an even number of minus signs is the key; consequently a result can be identified.

Using the above solving procedure, a transformation is defined by solving the four equations in the table, $a_1b_1 = -1$, $a_1b_2 = 1$, $a_2b_1 = -1$, $a_2b_2 = 1$, to give $\underline{\mathbf{a}} = [1\ 1]'$ and $\underline{\mathbf{b}} = [-1\ 1]'$. The probabilities can be h-transformed as above to show $\underline{\mathbf{a}\ \mathbf{b}} = -\underline{\mathbf{0}\ \mathbf{1}}$. The minus sign will not be observed after sampling, so the output will be $\underline{\mathbf{0}\ \mathbf{1}}$ which happens to look like a count of binary 1. Counting from zero, position 1 is where the original 1 locates in the original Table for this function. So what satisfies the function has been found.

This method of transformation was inspired by *Grover's quantum algorithm* [4]. It works for more than two simulated qubits, but for $n > 1$ iteration is necessary, about $2^{n/2}$ iterations. To summarize, (1) apply the function to tag a prepared list, which in reality is filled with various fractions to achieve probability normalization; (2) apply $\mathbf{H}^{\otimes n}$ to the tagged list in order to reorganize it; (3) reverse the phase of the first term in the list; and (4) solve for an updated result $\underline{\mathbf{a}\ \mathbf{b}}$. This last step is equivalent to applying $\mathbf{H}^{\otimes n}$ again.

Repeat steps (1)–(4) roughly $2^{n/2}$ times. When the iterations are completed, sampling may take place to provide with good probability a solution to the problem of what satisfies the given function. The solution must be checked by substitution, and if it fails, the method is repeated.

This method is a vast improvement over having to evaluate a function for up to 2^n times (once for each possible input combination), trying to hit upon what satisfies it.

Increasing the Capacity of Long-Term Memory

Simulated qubit combinations might be useful for portions of long-term memory. For instance, using two simulated qubits, binary-like information may be created with combinations of [1 0]′, [0 1]′, and η [1 1]′, employed two at a time. Using [1 0]′, [0 1]′ the following probability vectors may be created: [1 0]′, [1 0]′; [1 0]′, [0 1]′; [0 1]′, [1 0]′; [0 1]′, [0 1]′, these being ordinary binary coding $\underline{\mathbf{0\,0}}, \underline{\mathbf{0\,1}}, \underline{\mathbf{1\,0}}, \underline{\mathbf{1\,1}}$ which can be sampled and read with certainty. In addition, using [1 0]′, η [1 1]′ the following additional codes may be created: [1 0]′, η [1 1]′; η [1 1]′ [1 0]′; using [0 1]′, η [1 1]′ the following additional codes may be created: [0 1]′, η [1 1]′; η [1 1]′, [0 1]′; and finally there is the code η [1 1]′, η [1 1]′. Components containing η [1 1]′ have to be read more than once to determine their presence. Assuming this is done, two simulated qubits can store 9 codes, which is more than $2^2 = 4$ codes. The improvement increases exponentially with an increase in n.

Generally the number of additional data items grows exponentially with the number n, the number of independent multivibrators. By considering independent elements as being [1 0]′, [0 1]′, and η [1 1]′ there could be $2^n + n\{2^{n-1} + 2^{n-2}, \cdots, 2^1\} + 1$ codes. This, the author's calculation, is the binary count 2^n of the basic variables $\underline{\mathbf{0}}$, $\underline{\mathbf{1}}$; plus the binary count with η [1 1]′ in place of one variable; plus the binary count with η [1 1]′ in place of two variables; and so on to η [1 1]′ in place of all n variables.

The net count using n simulated qubits is far more than the measly 2^n codes using n bits. However, since some of the codes are probabilistic, waveforms would have to be sampled several times in order to see faithfully the original data. But this is easily accomplished in a classical system by permitting several sampling pulses and letting the results accumulate in a special register.

Conclusions

This chapter introduces multivibrator functions on simulated qubits. Multivibrator functions are those which modify the position of a state vector with the end result that one or more probability combinations are tagged with negative signs. Such tags are invisible to a sampling system because sampling only indicates a positive probability for a given output combination. Sampling cannot determine phase.

Fundamental to multivibrator functions are h-transforms, introduced above, that permit an ability to distinguish the state η $[1\ 1]'$ from the state η $[1\ -1]'$ after sampling. With the help of such transforms it was shown that a given but unknown multivibrator function in the symmetric and antisymmetric class may be classified conveniently.

For one simulated qubit, $n = 1$, symmetric and antisymmetric includes constant functions with output fixed at 0 or 1, and nonconstant functions (inverters and buffers). In general, for symmetric and antisymmetric functions involving n simulated qubits, for $n > 0$, a single readout is enough to classify the multivibrator function (normally up to 2^n evaluations of a given but unknown binary function would be necessary).

A multivibrator function is identified to within a complement, since complements cannot be distinguished. For example, the functions "0110" and "1001" are complements, but give the same sampled readout using the above transforms.

An interesting concept is a transformation that will help to identify what satisfies a given multivibrator function. Of special interest is a decoding function. This is a multivibrator function in which one of the outcomes is tagged with a negative sign (in probability space).

If the multivibrator function is sophisticated, it will be far from obvious what satisfies it. For example, there are Boolean equations that serve to define the prime factors of a large integer. Using brute force, it could take many trials to discover what satisfies such equations, since a great many prime numbers would have to be tested. But if such a function is implemented as a multivibrator function on simulated qubits, what satisfies it may be found using relatively few operations.

A decoding function, after suitable transformation, needs to be evaluated roughly $2^{n/2}$ times. Ordinarily it would require up to 2^n evaluations to find what satisfies it. So there are significant advantages to multivibrator functions for larger n, when working on satisfiability problems.

A simulated qubit can encode a great deal of additional information in its frequency and phase. For example, by varying frequency, simulated qubits can store true, false, and true/false simultaneously. Advantages are easy to demonstrate on paper. Using binary combinations of [1 0], [0 1], and η [1 1], the number of codes that can be stored increases exponentially with n and is far more than a mere 2^n, which is all that n bits provide.

Memory is important, so options for increasing memory density cannot be dismissed without due consideration. The above multivibrator function identification methods and the above multivibrator function satisfiability methods are interesting to study, and they might be discovered to have a role to play biologically.

References

1. Burger JR (2011) Neural networks that emulate qubits. NeuroQuantology 9:910–916
2. Pittenger AO (1999) An introduction to quantum computing algorithms. Birkhauser, Boston, MA
3. Burger JR (2011) Symmetric and anti-symmetric quantum functions. arXiv:cs/0304016, 13 June 2011
4. Nielson MA, Chuang IL (2000) Quantum computation and quantum information, Cambridge series on information and the natural sciences. Cambridge University Press, Cambridge [Paperback]

Self-Study Exercises

(If you desire to keep a record of your efforts, please show a couple of steps leading to the answers.)

The Phase of the 1

1. What is observed after sampling?

   (a) η [1 1]′ (ANS. True with 50 % chance; false with 50 % chance)
   (b) η [1 −1]′ (ANS. True with 50 % chance; false with 50 % chance)

2. What is observed for the tagged list η [−1 1]′ using transforms as in the text to determine its properties after a readout? (ANS. The transform gives [0 −1]′ but the negative sign is not seen. One reads out a true. So this is the result of a nonconstant binary function, an inverter)

Symmetric and Antisymmetric Function Determination

3. List a truth table for the following:

   (a) NOT function in shorthand form. (ANS. 1 0; means as the input counts in binary up 0 1 the output follows as 1 0)
   (b) All symmetric and antisymmetric functions for $n = 2$, in shorthand form. Place them in numerical order (there are 2×2^n or 8 functions and so there are 8 truth tables: "0000, 0011, 0101, 0110, 1001, 1010, 1100, 1111").
   (c) Label the above functions to be symmetric or antisymmetric, and regular or complementary (positive or negative). (ANS. s, a, a, s, s, a, a, s; +, +, +,+, −, −, −, −)

4. Apply the procedure given in the text to identify the following function: "1001" when assumed applied to a multivibrator. (ANS. $a_1 = -1$; $a_2 = 1$; $b_1 = 1$, $b_2 = -1$ which would be observed as **1 1**)
5. Can you discover another way to express the last transformation that identifies a symmetric and antisymmetric function? (HINT: Multiply the tagged list $[-1\ 1\ 1\ -1]'$ by $\mathbf{H}^{\otimes n}$ with $n = 2$.)
6. What is the potential savings in effort for symmetric-antisymmetric function identification when comparing simulated qubit versus classical function identification assuming $n = 10$? (Roughly 1 evaluation of the function versus up to $2^{10} = 1{,}024$ evaluations)
   (a) Why might classical identification be less than 2^n? (ANS. Symmetry might be used to advantage)

Satisfiability

7. Assume a truth table as follows: "0001." Trace through the listing method given in the text to find what satisfies this function. (ANS. $a_1 = -1$, $a_2 = 1$, $b_1 = 1$, $b_2 = -1$ or what is read is **a b** = **1 1**)

Data Packing

8. Use combinations of $[1\ 0]'$, $[0\ 1]'$, and $\eta^\circ = \eta\ [1\ 1]'$ assuming three simulated qubits:
   (a) Count with these codes (ANS. 0, 0, 0; … 1, 1, 1; … $\eta^\circ, \eta^\circ, \eta^\circ$)
   (b) How many codes are there? (ANS. $2^n + n\{2^{n-1} + 2^{n-2}, \cdots, 2^1\} + 1 = 29$)
9. Devise a way to detect $\eta\ [1\ 1]'$. (HINT: Several qubit readouts are necessary.)

Chapter 11
Postscript

The Search for the Source of Human Intelligence

There has always been a search for deeper computational power within the brain, needed for not only higher intelligence but also common sense, truth judgment, intuition, and artistic appraisal. Are computers with conventional gates capable of delivering inspired intelligence? Not according to several psychological and philosophical writers [1, 2] who seldom miss an opportunity to point out that the brain could not function as it does using conventional gates, and that artificial intelligence using computers as we know them is futile and doomed to failure.

But what if there were a much wider variety of logic? What if it were connected in new and different ways? And what if there were incomprehensible numbers, hundreds of trillions ($\sim 10^{12}$) of gates, instead of just millions ($\sim 10^{6}$) as in current computers? Under these conditions skeptics might have to reevaluate a brain's computational possibilities.

Energized by books read as a youth [3], the author has revealed that neurons do indeed provide a wider variety of logic, such as pulsed dendritic logic gates, enabled logic gates, regular and weak synapses to send signals where they are needed, and short-term memory neurons to serve as timers. The choice of gates is very much wider than what is possible using solid state, which tends to be limited to direct current logic, that is, false is one voltage, and true is another. This style of logic generally takes excessive energy to switch from one state to another. Neural pulses, in contrast, avoid voltage levels, and are likely more efficient with power and energy.

Pulsating logic better supports an asynchronous system, since it avoids the need for a central clock as used in desktop computers. Unfortunately, high-speed clocking requires a lot of power because of incessant high current spikes and because clocks actually make parallel processing more difficult. Clocking suffers from the need to employ synchronous registers everywhere, and clock skew, where slow-moving signals as in neurons lose synchronization.

J.R. Burger, *Brain Theory From A Circuits And Systems Perspective*, Springer Series in Cognitive and Neural Systems 6, DOI 10.1007/978-1-4614-6412-9_11,
© Springer Science+Business Media New York 2013

As if the count of brain neurons were not high enough, there continues a search for deeper computational power inside the neuron, hoping to find logical capabilities among molecules and electrons. An interesting proposal explored in this book is that electrons are available in postsynaptic regions, and indeed are necessary to synchronize presynaptic actions, not just for one synapse, but for all of them [4]. Another interesting proposal considers subneural quantum computing [5]. Lacking scientific verification of an interneural quantum system, this book has suggested the prescientific possibility that internal molecules may actually serve as simulated qubits, doing much but not all of what quantum particles do.

Much has been written about the meaning of consciousness, that is, subjective consciousness that is limited to the creature that experiences it. However this book tends to avoid consciousness as a philosophical issue. Instead this book simply equates consciousness to short-term memory, that is, common persistent memory that permits one to remember briefly sensory images, sights, sounds, and recalls. The focus of this book is on *stream of consciousness*, a deep-rooted concept developed by such authors as James Joyce and psychologist William James.

Stream of consciousness is a result of sensory inputs and associative memory recalls as laid out in this book. In order to synthesize a neural system, a goal of this book, it was absolutely necessary to go beyond current dogma concerning circuits in general and neural networks in particular. For instance, circuits are not limited to networks of wire as circuit societies prefer to emphasize, and neural networks are not limited to weighted synapses and activation functions, which are merely a limited form of neural logic known as enabled logic in this book.

The "Neural Circuits and Neural Systems" Point of View

Obviously the brain is more than a collection of molecules, although one might think otherwise given the current popularity of molecular biology and biochemistry. This book views the brain as a connected system of neurons, not just a simple system that is spoken of in the abstract, but a specific complex circuit of neurons whose purpose is to chain together and to purposefully blend a wide assortment of neural pulses. As in any electrical system, pulses are subject to the laws of circuits and systems. For instance, they propagate along a path at a given rate, transferring charge predictably. Overall, none of the established laws of physics can be violated by a brain, since it exists in the physical world.

Unfortunately, proposals occasionally ignore the laws of circuits and systems, or so it seems, possibly because the proposals are incomplete. For instance, it is technically amiss to postulate memory centers in dendrites, without saying exactly how an associative search brings these elements to consciousness. Equally unsuitable is failing to explain how an image is committed to memory at the instantaneous speeds that people commonly experience during memorization. The point is, rules apply, for instances, connectivity applies, meaning that every subsystem must have a path to appropriate parts of the overall system, and timing applies, meaning that brain events must occur within appropriate time frames.

Circuits and systems are not vague concepts; they use realistic elements; and they either function correctly or they do not. Unlike dreamy concepts, circuitry has the advantage that it may be simulated in a computer, or physically constructed by clever engineers. There are only so many ways a functioning system can be synthesized, given what is known about how neurons pulsate and perform Boolean logic. Based on what is known of neural logic, this book presents a particular circuit for stream of consciousness. This book's system is not the only possible system, just one of many that could be proposed.

This book does not pursue reverse engineering based on physical dissection of the brains of lower life forms. To arrive at a fitting hypothesis, this book applies logic theory using realistic circuit elements to develop a system, and then analyzes the resulting interconnection to verify its correctness. Neural circuits and neural systems (NCANS) is a form of circuit science, but is definitely not the bailiwick of circuit engineers, since engineering is mainly a for-profit activity mostly limited to hardware. So unless practical medical or robotic devices are involved, engineering interest is likely to be lukewarm. Consequently owing to the hardware bias of circuit theory, the subject of NCANS has been neglected for many decades. The good news is, there remains an opportunity for young scientists to realize its importance, and to revive it.

Novel Circuit Elements for an Efficient System

Certain elements are necessary to a neural system but these novel elements might not be universally expected, at least not by circuit engineers. For example, an interesting circuit element has been described as a *weak* synapse, meaning it triggers and transmits only one dendritic pulse, a sort of pulsed *mono stable multivibrator*. Particular subcircuits also select and transmit only the first pulse of a burst, effectively accomplishing the same function as a weak synapse. Single pulses have been observed in brain cells, and their purpose has long been a topic of speculation. This book proposes that single pulses are extremely important to system timing.

Single pulses are needed for an accurate sequencing of subcircuits, which would not be efficient otherwise. For example, multiple returns from long-term memory need to be individually catalogued for a priority determination, which is best done with a single pulse moving along a delay line, as in this book's *recall referee*. This process could be accomplished with neural equivalents of a pulse counter and binary decoder, but this would require many more neurons and a complex logic network. Taking advantage of single pulses is considerably more efficient.

Single pulses are key to recursive neurons that recycle a single pulse, thus establishing an element of memory of a certain type. But such neurons, termed *multivibrators*, are far more than elements of memory. By varying the frequency of the cycling via changing loop delay, and then sampling the waveform with a single pulse, they take on some of the properties of quantum qubits. They become *simulated qubits*.

Simulated qubits are not capable of teleportation, since they are basically classical (non-quantum) devices. But like qubits, they hold true and false simultaneously with respective probabilities in probability space, where coordinates may be transformed by modifying the frequency and phase of the multivibrator waveform. When multivibrator functions are included as defined in a chapter of this book, simulated qubits possess interesting possibilities for function identification and function satisfiability. Without simulated qubits, these operations are difficult or impossible.

Recursive neurons operating as simulated qubits are better than ordinary logic for certain purposes, for example, for a *cue editor* which randomly removes cues to achieve a match during a drawn out memory search, as described in this book.

The most important function of simulated qubits is perhaps not probabilistic logic, but toggling, which can be accomplished with fairly simple non-sampled circuits as given in this book. Recursive neurons can be made to toggle between true and false with a single pulse acting as a trigger. This type of element was shown to be quite helpful in achieving multi write and multi read capability in human associative memory. Furthermore, arrays of controlled toggles operating in parallel may accomplish rapid mental arithmetic, all subliminal, for the purposes of a quick computation of the priority of a return. This helps assure that only important recalls are permitted into conscious short-term memory.

The power of reversible parallel arithmetic using controlled neural toggles cannot be stressed enough. It has been proposed that controlled neural toggles are the secret weapon used by savants who perform amazing mental calculations, such as multiplying large numbers, or identifying large prime numbers.

Toggle arithmetic is accomplished when a set of *source* toggles control a disjoint set of *destination* toggles. If all of the source toggles have true outputs, then all of the destination toggles are made to change their states, or flip true to false, or false to true. Arithmetic is accomplished by a meaningful sequence of such operations. This approach is logically reversible, and may be physically reversible as well, which implies energy efficiency.

There are situations under which toggles are unconditionally reset to zero, and this of course is not reversible. Resetting effectively occurs, for example, when cues in a cue editor are replaced, or when accumulated priorities in a recall editor are overwritten at the beginning of a priority calculation.

Another interesting circuit element is the *short-term memory neuron*. To work within a system, this neuron must emit regular pulses that can be recognized by connected neurons. The explanation underlying its operation is that an extended dendritic pulse occurs when its membrane is charged, but is not permitted to discharge normally. Charge is held on the inside surface of the membrane relative to the outside. A short-term memory membrane discharges at a slow controlled rate, long enough to accomplish an extended burst of regular pulses, depending on internal ionic conditions and on conductance to the external surface.

To have a long dendritic pulse, and a long trigger at the soma, which implies a long burst of regular pulses at the axon, conductance must be lower than what it is for a membrane directly exposed to the ionic solutions. This suggests regions of myelination or neurotransmitter involvement on the surface of the dendrite, to insulate it, as is typical for axons.

Short-term memory neurons, by the way, are not just for consciousness; they also help regulate the asynchronous system. For instance, a short-term memory neuron conveniently serves to extend in time the duration of a regular pulse burst. This occurs, for instance, when a long-term memory element must remain active longer than its normal ten or so pulses, to ensure that it is read properly. This application has been termed *burst stretching*.

Another type of application occurs when a short-term memory neuron is used as a *timer* for a cue editor so that, after a given duration with no results, the timer indicates that a memory search has failed to find matches.

Perspective on Synapses

Synapses are very numerous, averaging about 500 per neuron for adults, but they are not all receiving signals concurrently from other neurons. They are available, however, and easy to find by interneurons that need connections to facilitate a phenomenon known as neural placidity. For modeling purposes, regular (fast) excitatory synapses often can be represented by a transconductance amplifier giving positive current pulses, while regular (fast) inhibitory synapses often can be represented by a transconductance amplifier giving negative pulses. Transconductance amplifiers effectively provide a packet of charge when a given neural voltage pulse is applied to the input.

Transconductance amplifiers are appropriate models because they tend to operate "one way", as does a synapse, from presynapse to postsynapse; they result in a trigger by charging membrane capacitance. They avoid unrealistic current spikes and unworkable loading.

Synaptic receptors often occur on spines, although other locations are possible. When a presynaptic vesicle releases its excitatory neurotransmitters, they interact with the postsynaptic neural membrane so as to trigger it. A membrane with typical properties generates a regular neural pulse, which is a certain waveform between about −70 and +40 mV, composed of pulses about 1 ms wide. Synapses are not weighted in this book, as they generally are in the field of artificial neural networks; their only role is to trigger standard pulses.

Neurotransmitters are released in the cleft. Intuitively, positively charged excitatory neurotransmitter ions can be imagined as being repelled away from the receptor into the thin synaptic cleft by the first positively charged neural pulse. But they are soon attracted back to trigger additional pulses. The resulting burst is assumed terminated only when the presynaptic bouton returns to its negative rest potential; this in turn attracts home with electrostatic force all positive neurotransmitters. Back propagating pulses from a soma are quite common and also tend to encourage positively charged neurotransmitters to leave the postsynaptic regions.

As pulses propagate, pulse shape remains more or less regular, similar to the soliton of electrical theory. Regular pulse waveforms are expected in exposed membrane that is uniform and continuous. Propagating pulses undergo dendritic logic and also enabled logic, where enabled logic is more akin to what is accomplished by the activation functions in the engineering field of artificial neural networks.

Plethora of Types of Neural Logic

Dendritic logic was shown in this book to depend on the timing of pulses propagating along dendritic branches, where pulses must arrive concurrently to accomplish the desired logic, which may be AND, OR, XOR gates; note that NOT gates may be derived from the XOR (exclusive OR) gates [6]. Simulations suggest that membrane activity and geometry at a dendritic junction determine the resulting logic. Similar logic might very well be generated by an inhomogeneous region of a dendrite, although appropriate geometry is more difficult to envision. Thus the dendrites of a neuron, when they are extensive, may support thousands of logic gates.

In stark contrast, enabled logic was shown in this book to depend on a smaller dendrite making contact to a larger body or soma. The resulting added capacitance, when hit by pulses, thus accumulates voltage and eventually reaches a threshold for triggering. Voltage is accumulated from standard neural pulses (not weighted synapses) that add to produce AND, OR gates.

Enabled NOT gates are theoretically possible, but only when inhibitory neurotransmitters are considered. The XOR function is not directly available in a single node. The XOR may be generated using two layers of enabled logic with some sort of complex weighting to realize the equivalent of negative weighting factors. Alternatively, simply employ three layers using AND, OR, and NOT to realize the Boolean expansion of the XOR function.

Enabled logic is not as dependent on timing as dendritic logic, so it has a distinctive place in neural circuitry. Enabled logic is generally limited to neural regions that can store adequate charge, generally the soma. This suggests only one enabled gate per neuron. In contrast, dendritic gates may occur by the thousands in a given neuron.

Brain System Requirements

What a brain does is based on common experience: Obviously sensory impressions are projected into conscious short-term memory. Interspersed with these are recalls from associative long-term memory. Recalls are extremely important to intelligence, because they provide reactions to the environment, and they bring forth secondary recalls that suggest reactions.

Reactions are important to survival. If there is a memory block in the face of danger, the consequences are catastrophic. A great life-saving feature is cue editing to fight a tendency for mental blocks, especially when it is really important to remember what has happened in past similar situations, for instance, when petting an angry dog.

Even so, it often happens that a person forgets something, even though there are many cues. A *cue editor*, as given in this book, will work subliminally to remember a forgotten thing. It sometimes pops into one's head hours later, indicating that a cue editor has been at work subliminally.

Just as bad for survival are multiple recalls that confuse and effectively paralyze a person. It is critically important to survival to see only important recalls concerning something relevant, as opposed to experiencing pictures of red apples, red race cars, or red faces when in fact there is a red fire in your bed. Decisive action can be critical for self-preservation.

For survival, and fundamental to any system of associative memory, human memory included, is multiple match resolution. First of all, it must prevent basic memory breakdown because of multiple returns on the same neural paths. Secondly, a basic *recall referee* is essential, including a processor that assigns priority to memory returns and then chooses the highest priority image to be gated into conscious short-term memory. A basic recall referee given in this book is based on accurate digital principles and avoids error-prone analog processing.

Those familiar with associative memory structures will realize immediately that short-term and long-term memory must work together intimately, as in this book, with the aid of well-oiled editors for memory search, to prevent mental blocks, and recall referees, or editors, that choose important recalls based on quick calculations of image priority. This book provides not just ideas and concepts but also physical circuits that, with proper care, would work as a neural system to achieve stream of consciousness.

Conclusions

Can hundreds of billions of neurons and hundreds of trillions of synapses carry out what a human brain accomplishes? This remains to be seen. Indeed, there are times when even the author doubts that classical logic explains all that a brain does, especially after hearing about the underwhelming performances of the so-called intelligent robots and computers. It seems that there has got to be more to intelligence than merely imitating it with a clever computer program.

This book considers systems radically different from those being tried in robotics and artificial intelligence, where an immediate payoff is usually demanded. First off, this book considers a plausible system of neural logic based on realistic neural gates and multivibrators using recursive neurons. Simulated qubits and controlled toggles are explained while keeping in mind the possibility of physical qubits, and at times, even electrons, which are hardly ever mentioned in neuroscience.

The study of brain systems is embryonic and occasionally mystifying, but there are many possibilities and opportunities. This book has revealed a vision of a system with its own stream of consciousness, using principles well known in the field of circuits and systems. It seems certain that, if nothing else, the conscious brain is in fact a complex electrical system, and is best modeled as an electrical system.

References

1. Schacter DL (1996) Searching for memory: the brain, the mind, and the past. Basic Books, New York
2. Penrose R (1989) The emperor's new mind: concerning computers, minds and the laws of physics. Oxford University Press, Oxford
3. Berkeley EC (1949) Giant brains; or, machines that think. Science Editions, Inc., New York
4. Walker EH (2000) The physics of consciousness: the quantum mind and the meaning of life. Perseus, Cambridge, MA
5. Hameroff S (2007) Orchestrated reduction of quantum coherence in brain microtubules: a model for consciousness. NeuroQuantology 5:1–8
6. Burger JR (2010) XOR at a single point where dendrites merge. arXiv:1004.2280 [cs.NE] 13 Apr 2010

Appendix A

Simulation Experiments for Neural Pulses

Introduction

Electrical Parameters

Parameters may be estimated from standard values. For these experiments parameters are assumed as in Table A.1.

A simplified description of a neural pulse is as follows: Initially the voltage across the membrane is −70 mV inside relative to outside. A pulse is triggered by increasing inside voltage up to about −55 mV. Triggering in a simplified model means that sodium currents are flowing in and that potassium currents are flowing out, at the same time. Potassium currents are much less mainly because of the physical properties of potassium, resulting in a buildup of positive charge inside the membrane.

Currents charge membrane capacitance, 1 μF/cm^2 which translates into a certain internal potential. When internal potential reaches the sodium cutoff voltage at about +40 mV, sodium current cuts off, leaving potassium current to discharge membrane capacitance. When internal voltage has dropped to the potassium cutoff voltage of −90 mV, potassium current cuts off. Once current sources are removed, the internal voltage is assumed to drift via membrane conductance (0.3 mS/cm^2 or 3.333 kΩ/cm^2) up to its rest value of −70 mV.

Generally a pulse may be approximated by straight line segments as in Fig. A.1. An actual pulse would have rounded tips because the cutoff voltages vary slightly from molecule to molecule in a membrane. The lines in an actual pulse would not be

J.R. Burger, *Brain Theory From A Circuits And Systems Perspective*, Springer Series in Cognitive and Neural Systems 6, DOI 10.1007/978-1-4614-6412-9,
© Springer Science+Business Media New York 2013

Table A.1 Simulation parameters

Rest potential	−70 mV
Trigger voltage (both Na and K currents)	−55 mV
Na current cutoff voltage	+40 mV
K current cutoff voltage	−90 mV
Sodium current	67.25 μA/cm^2
Potassium current	30.4 μA/cm^2
Membrane capacitance	1 μF/cm^2
Membrane conductance	0.3 mS/cm^2
Internal resistivity ρ	15.7 Ω cm

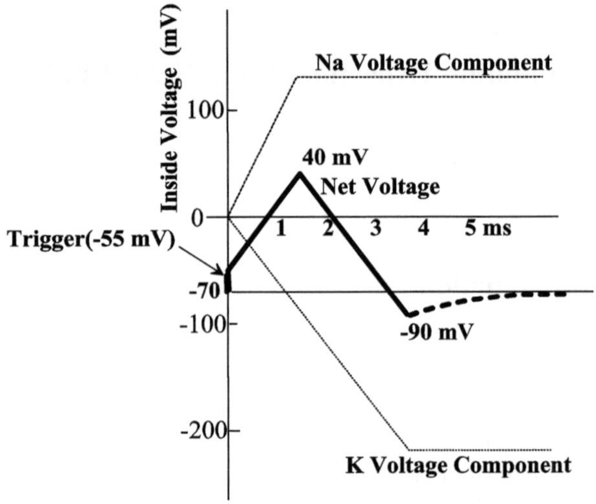

Fig. A.1 Simplified hand analysis of a neural pulse in terms of Na and K currents

perfectly straight because of conductance effects that were ignored in the figure. However, although an approximation, much can be learned from a simplified model.

Next the above parameters will to be scaled to the physical sizes of the regions involved.

Physical Parameters

Short lengths of dendrite can be modeled as being tubular as in Fig. A.2.

Segments could be any convenient size in a linear model, but here each segment has a fixed diameter $D = 1 \times 10^{-4}$ cm and length $L = 0.1$ cm. The number of segments n is left variable. Table A.2 summarizes equations for computing electrical parameters from physical parameters. Table A.2 summarizes the results of a calculation for the given segment.

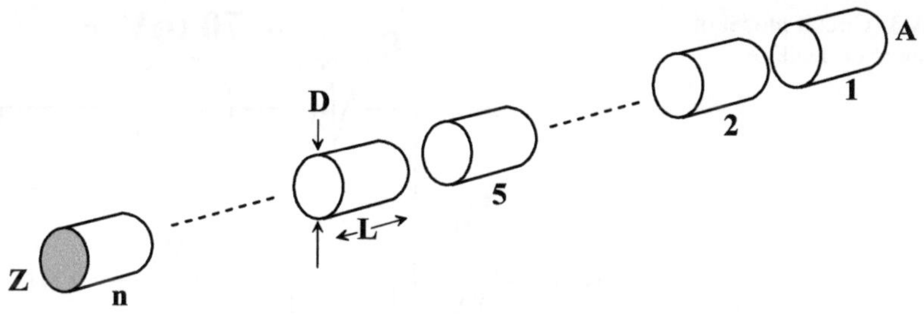

Fig. A.2 Segments of dendrite for modeling purposes

Table A.2 Equations

$L = 1$ mm $= 0.1$ cm
$D = 1\ u = 1.0\text{e}{-4}$ cm
$A_{CS} = \pi(D/2)^2$
$R_1 = \rho L/A_{CS} = 15.7L/A_{CS}$
$A_{SIDE} = \pi DL$
$C_1 = 1\ \mu\text{F/cm}^2 \times A_{SIDE}$
$R_{LOSS1} = 1/(0.3\text{e}{-3} \times A_{SIDE})$
$I_{Na1} = 0.\ 06725\ \text{mA/cm}^2 \times A_{SIDE}$
$I_K = 0.0304\ \text{mA/cm}^2 \times A_{SIDE}$

Simulation Tools

Simulations are performed with a simple but reliable and transparent circuit analysis program such as Winspice, available inexpensively for download (http://winspice.com; accessed March 4, 2013). The currents in the above list are controlled current sources denoted by G. Triggering at an internal voltage of about −55 mV activates both sodium and potassium currents. The sodium current must cut off and stay off when internal voltage reaches 40 mV. The potassium current remains until voltage drops to −90 mV, at which point it also must cut off and stay off. The electrical circuit of a segment can be modeled as in Fig. A.3.

Parameters are as in Table A.3. The current sources depend on internal voltage v_n, as suggested by the simplified neural pulse. The entire segment is referenced to a rest voltage of −70 mV. This condition is modeled by placing −70 mV in series with the conductive loss resistance R_{Loss1}. Voltages on the capacitance and current sources do not matter and are simply tied to a "ground" reference, which means zero voltage. The suggested code for a subcircuit representing a segment is as follows.

```
.subckt seg1 1 2
vNernst 3 0 −70 mV
rNernst 3 1 106MEG
cseg 1 0 31.4p ic = −70 mV
rseg 1 2 200MEG
xcur 1 0 cur1
.ends
```

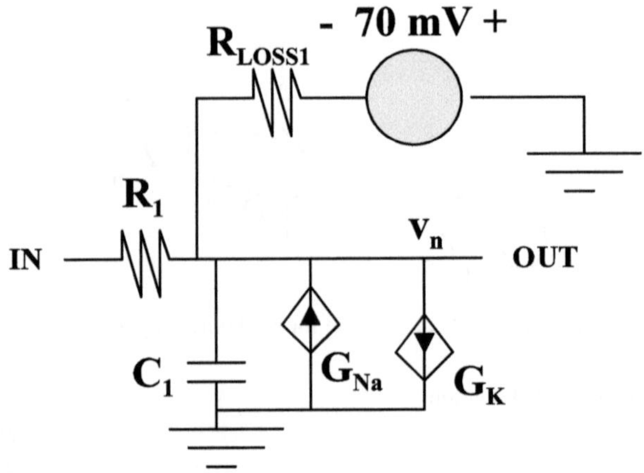

Fig. A.3 Circuit model of a segment of dendrite

Table A.3 Electrical parameters

$R_1 = 200$ M
$C_1 = 31.4$ pF
$R_{\mathrm{LOSS1}} = 106$ M
$I_{\mathrm{Na1}} = 2.11$ nA
$I_{\mathrm{K1}} = 0.955$ nA

xcur refers to a subcircuit **cur1** that provides the appropriate charging and discharging currents, as explained below.

Controlled Current Source Subcircuit

The current sources in the given code are represented by a subcircuit termed **cur1**. Currents are arranged to switch at given voltage values. To accomplish this, node voltage v_n must be sampled. This requires care since node resistances in the model are high, hundreds of megohms ($10^6\ \Omega$). Sampling will not disturb the node if it is done through tens of gigaohms ($10^9\ \Omega$). In order to control the current sources, three control voltages v_1, v_2, v_3 are going to be created. The currents will be delivered by voltage-controlled current sources. These control voltages will depend on node voltage v_n.

Starting Sodium Current To begin, sodium current switches on as node voltage increases through about −55 mV. Figure A.4 illustrates the output of a basic nonlinear amplifier or comparator whose purpose is to generate a control voltage v_1.

A schematic for the amplifier appears in Fig. A.5. Note that the reference voltage is **−55 mV** and there is no feedback resistor.

The amplifier is modeled with a voltage-controlled voltage source E that amplifies by a factor of 100 the voltage across the two (+, −) inputs to the voltage source. Ideal diodes are used to clip the output to be between −1, +1 V. This is

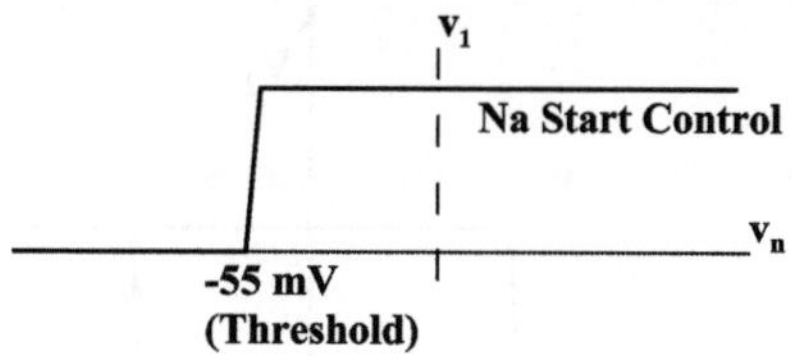

Fig. A.4 A control voltage v_1 to start sodium current

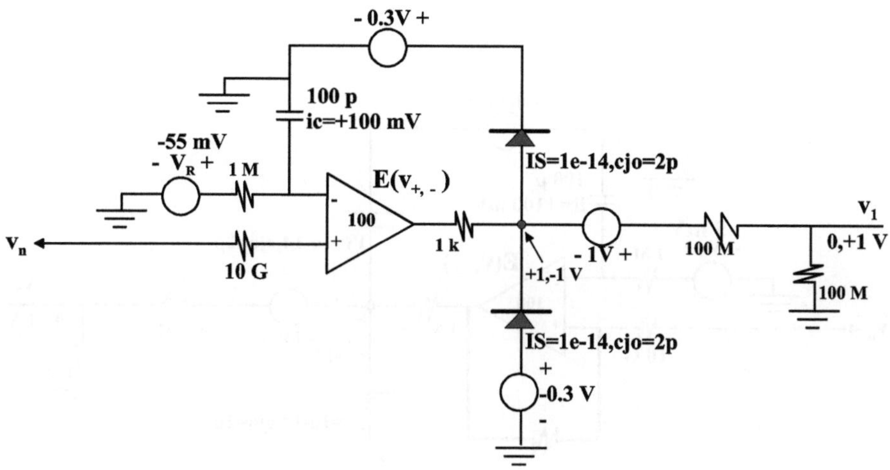

Fig. A.5 Amplifier to generate v_1

translated through a floating fixed source of 1 V to be 0, +1 V. These are the output levels for v_1.

A reference voltage V_R to the "−" input determines the threshold, $V_R = -55$ mV. One hundred picofarads (pF) of capacitance, or 100×10^{-12} F, with an initial condition of 100 mV assures that the amplifier forces v_1 to 0 at $t = 0$. Not shown is a small capacitance of 1 femtofarad (fF) or 10^{-15} F, which is included from the "+" input to ground to ensure numerical stability.

Stopping Sodium Current Stopping the sodium current at about +40 mV requires a voltage v_2 that goes true at +40 mV, but which stays true until node voltage is safely below the threshold for triggering at about −55 mV. This sort of behavior may be accomplished with a nonlinear amplifier that generates a hysteresis loop. The desired loop appears in Fig. A.6.

A circuit that provides this loop is shown in Fig. A.7. Note that the reference voltage is −5 mV and the feedback resistor $R_f = 150$ G.

The sodium current is stopped by a simple expedient, making control voltage v_1 zero by switching it to ground with a controlled switch as in Fig. A.8. Switch S2 is a voltage-controlled switch that closes when v_2 goes true or high.

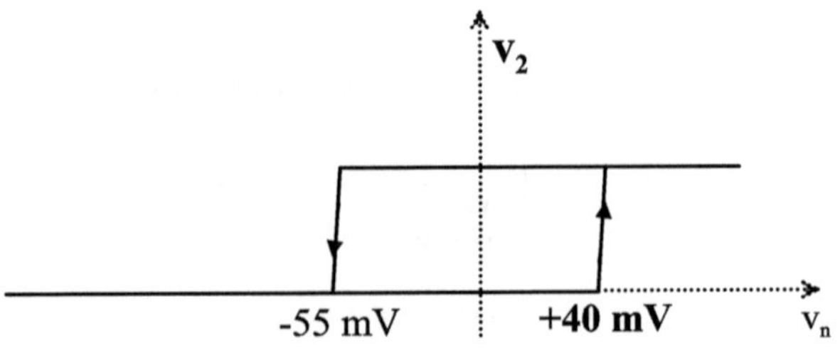

Fig. A.6 A control voltage v_2 to stop sodium current

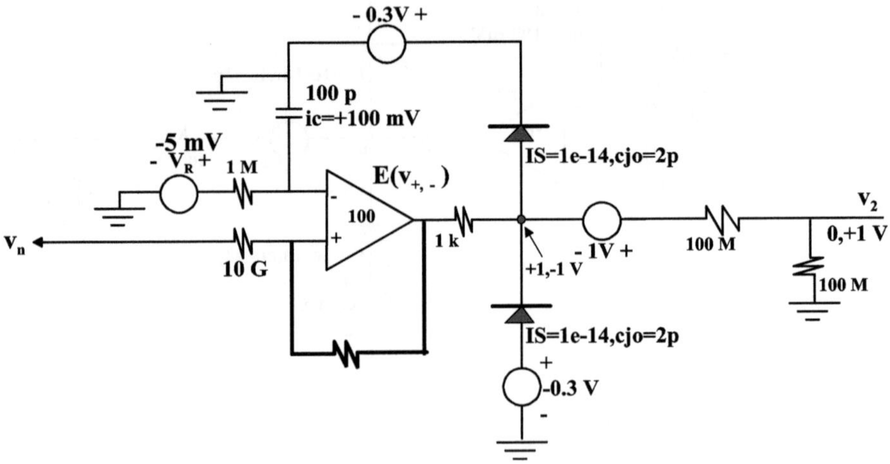

Fig. A.7 Hysteresis generating amplifier to provide control voltage v_2

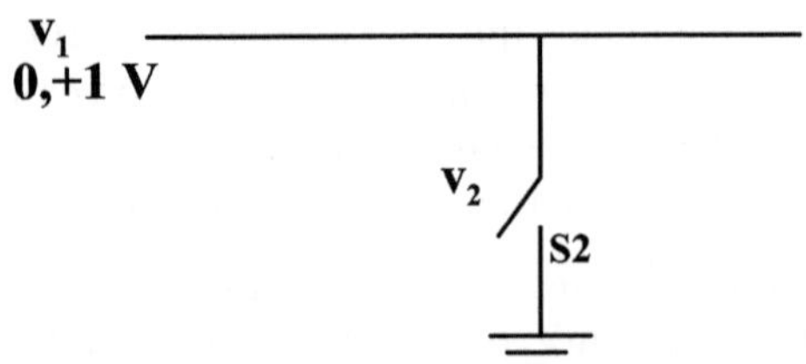

Fig. A.8 Sodium current stop switch

Potassium Current Potassium current is modeled as beginning at the trigger threshold of about −55 mV and stopping at −90 mV. Control voltage v_3 may be generated with a hysteresis loop as in Fig. A.9.

The circuit for this is shown in Fig. A.10. Note that the reference voltage is −70 mV and the feedback resistor $R_f = 300$ G.

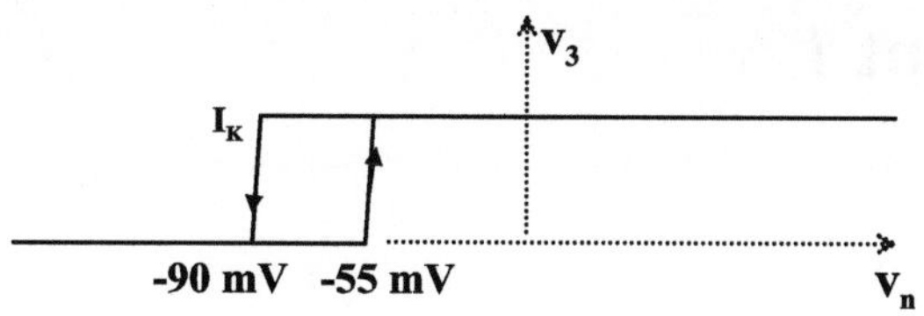

Fig. A.9 Control voltage v_3 to start and stop potassium current

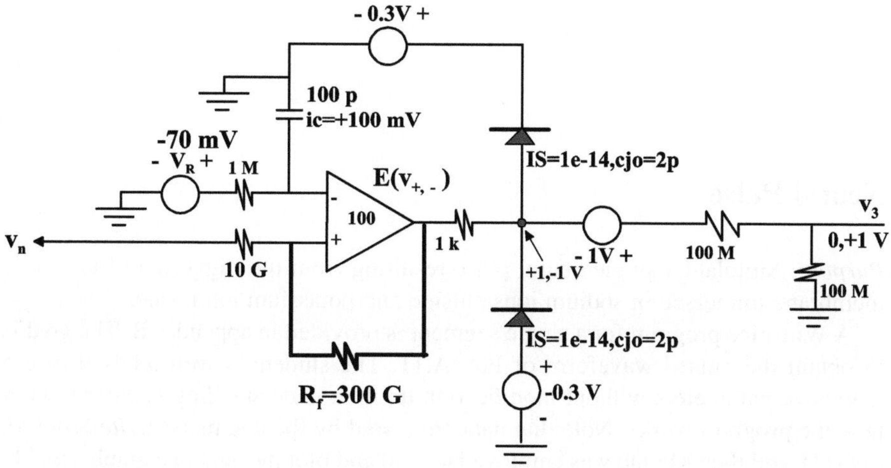

Fig. A.10 Circuit for control voltage v_3

Typical Winspice code is provided in appendix B. Eventually differing segments are going to be connected together in various ways. When different parameters appear in a segment because of inhibitory neurotransmitters or myelination, as in future experiments, a new subcircuit other than seg1 must be created for each case.

Experiment 1

Neural Pulse

Purpose: Simulate a single neural pulse resulting from the triggering of an active membrane immersed in sodium ions outside and potassium ion inside.

A Winspice program for a single segment is provided in appendix B. The goal is to obtain the pulsed waveform of Fig. A.11. The student is well advised to use the above parameters without modification until an understanding is gained about how the program works. Note that data are stored by the command *write SingPul. csv v(1)*, and then Matlab was employed to read and plot the data in a graph suitable for publication.

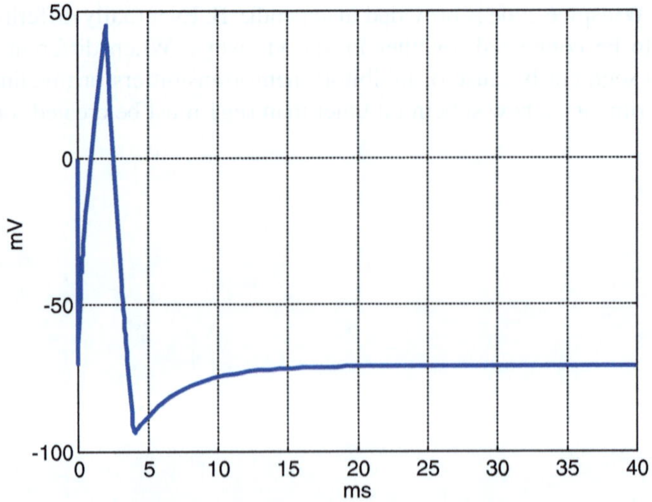

Fig. A.11 Single Pulse (SingPul)

- Show in a plot how sodium current can be made to double (slow) the rise time of the pulse (−55 to +40 mV).
- Show in a plot how potassium current can be made to double (slow) the fall time of the pulse.
- What is the minimum current amplitude of a trigger pulse?
  Note triggering is accomplished at a nominal 5 nA with:
  itrig 0 1 dc = 0 pwl(0,0,10us,5n,0.2 ms,5n,0.21 ms,0)
- Restore parameters that were changed.

Experiment 2

Pulse Propagation

Purpose: The objective is to estimate the speed of a pulse in a given dendritic tube. A dendritic tube can be modeled as follows.

```
x1 1 2 seg1
x2 2 3 seg1
x3 3 4 seg1
x4 4 5 seg1
x5 5 6 seg1
x6 6 7 seg1
x7 7 8 seg1
x8 8 9 seg1
x9 9 10 seg1
x10 10 11 seg1
```

Simulation results are in Fig. A.12.

- What is the speed of propagation in cm/s?
- Is there a back propagation from the end? How can you know?
- How much must segment capacitance be increased to increase delay (slow down) by 20 %?

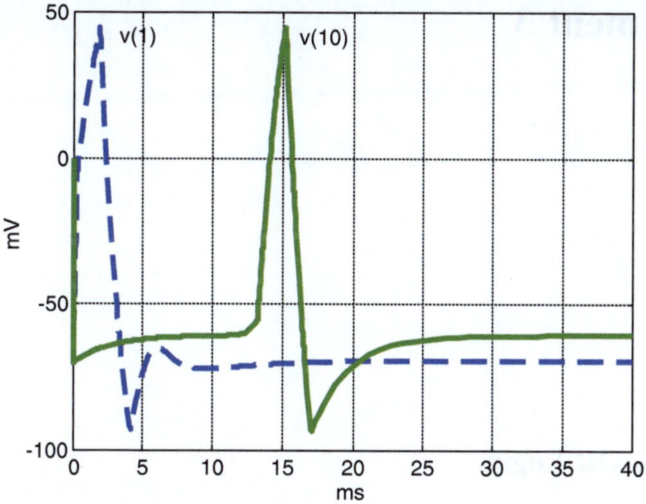

Fig. A.12 Propagation (Propag)

Experiment 3

Dendritic OR Logic

Purpose: The purpose is to demonstrate an OR gate using intersecting dendritic tubes.

The following code may be helpful.

```
itrig 0 1 dc = 0 pwl (0,0,5us,10n,.2 ms,5n,.21 ms,0)
itrig1 0 21 dc = 0 pwl (0,0,5us,10n,.2 ms,5n,.21 ms,0)
x1 1 2 seg1
x2 2 3 seg1
x3 3 4 seg1
x4 4 5 seg1
*
x21 21 22 seg1
x22 22 23 seg1
x23 23 24 seg1
x24 24 6 seg1
*
x5 5 6 seg1
x6 6 7 seg1
x7 7 8 seg1
x8 8 9 seg1
x9 9 10 seg1
x10 10 11 seg1
```

- How do you know in Fig. A.13 that there was a back propagation to v22? What caused it?
- How do you know in Fig. A.14 that there was a back propagation to v1? What do you suppose caused it?

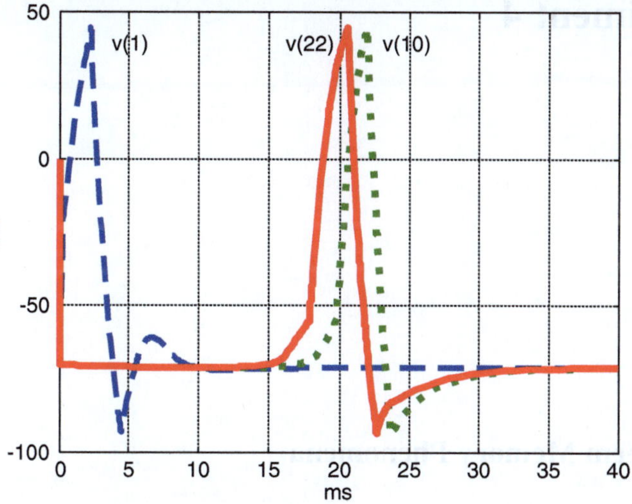

Fig. A.13 OR1 Note signal to v(22) from intersection

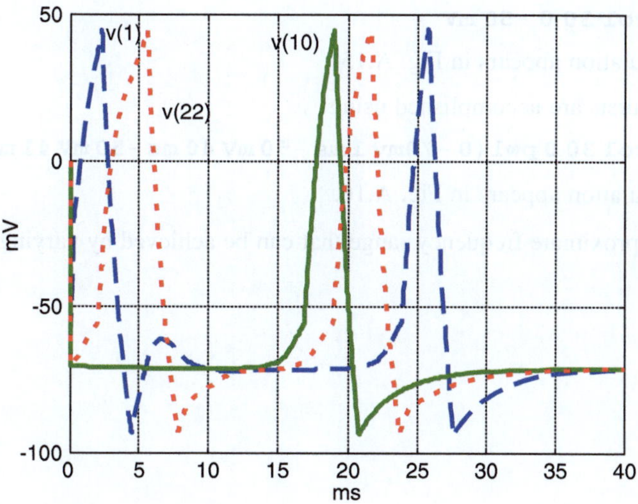

Fig. A.14 OR2 Note back propagations

Experiment 4

Short-Term Memory Phenomena

Purpose: The purpose is to demonstrate an extended burst of pulses interpreted psychologically to be short-term memory.

Note that charge storage in a dendrite is modeled by changing the voltage reference of a segment from −70 mV to some higher voltage above the trigger threshold of −55 mV; for example:

```
vcontrol 30 0 -50 mV
```

The simulation appears in Fig. A.15.

Timed bursts are accomplished using:

```
vcontrol 30 0 pwl (0 -70mv 1 ms -50 mV 40 ms -50 mV 41 ms -70 mV)
```

The simulation appears in Fig. A.16.

- What approximate frequency range that can be achieved by varying vcontrol?

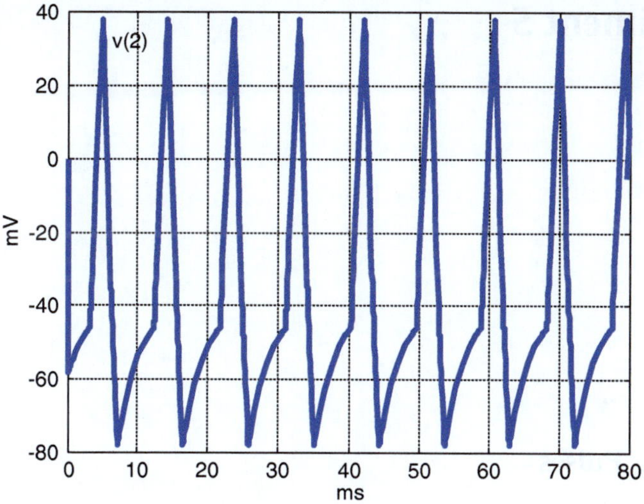

Fig. A.15 Continuous triggering representing memory

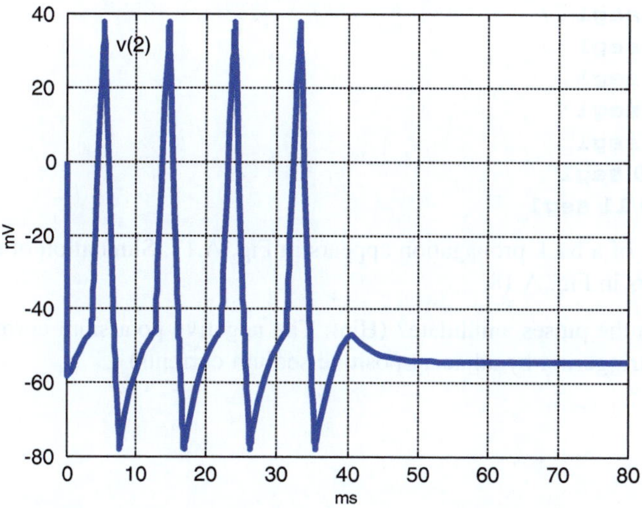

Fig. A.16 Timed Burst -50 mV for 40 ms

Experiment 5

Colliding Pulses

Purpose: The purpose is to demonstrate that colliding pulses annihilate each other. The following code may be helpful.

```
itrig 0 1 dc = 0 pwl (0,0,5us,10n,.2 ms,5n,.21 ms,0)
itrig1 0 10 dc = 0 pwl (0,0,5us,10n,.2 ms,5n,.21 ms,0)
x1 1 2 seg1
x2 2 3 seg1
x3 3 4 seg1
x4 4 5 seg1
x5 5 6 seg1
x6 6 7 seg1
x7 7 8 seg1
x8 8 9 seg1
x9 9 10 seg1
x10 10 11 seg1
```

Simulation of a back propagation appears in Fig. A.17. Simulation of an annihilation appears in Fig. A.18.

- Why do the pulses annihilate? (Hint: The negative potassium currents prevent further triggering by adjacent positive sodium currents)

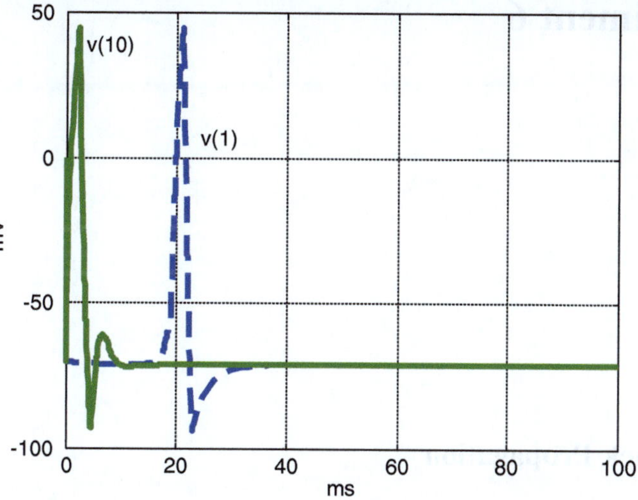

Fig. A.17 There is no v(1) pulse, just a v(10) pulse. Pulse moving backwards from v(10) arrives at v(1) at about 20 ms

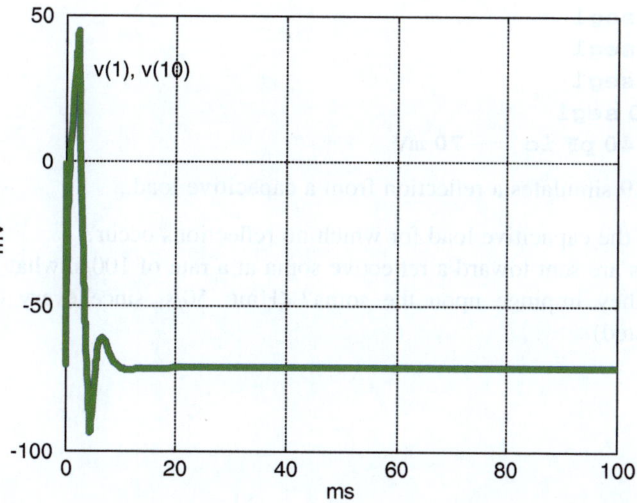

Fig. A.18 There are both v(10) and v(1) pulses. They collide and annihilate, resulting in no pulses at 20 ms

Experiment 6

Pulse Back Propagation

Purpose: The purpose is to demonstrate back propagation from a capacitive load. The following code may be helpful.

```
itrig 0 1 dc = 0 pwl (0,0,5us,10n,.2 ms,5n,.21 ms,0)
x1 1 2 seg1
x2 2 3 seg1
x3 3 4 seg1
x4 4 5 seg1
x5 5 6 seg1
x6 6 7 seg1
x7 7 8 seg1
x8 8 9 seg1
x9 9 10 seg1
c1 9 0 40 pF ic = −70 mV
```

Fig. A.19 simulates a reflection from a capacitive load.

- What is the capacitive load for which no reflections occur?
- If pulses are sent toward a reflective soma at a rate of 100/s, what is the rate at which they impinge upon the soma? (Hint: 50/s, since every other one is annihilated)

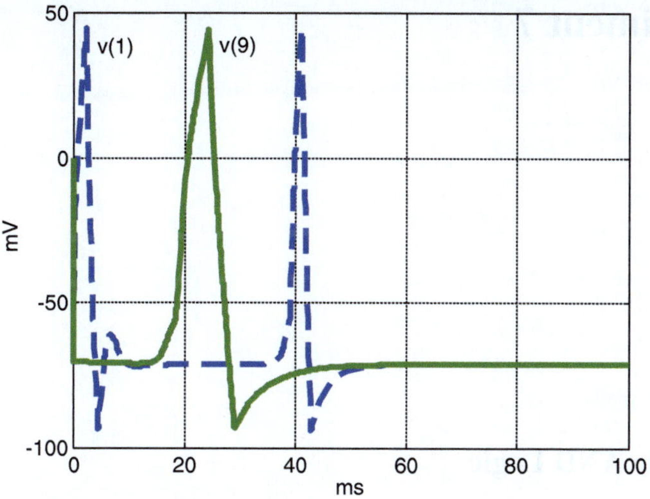

Fig. A.19 Reflection from 40 pF termination

Experiment 7

Dendritic AND Logic

Purpose: The purpose is to demonstrate AND logic for intersecting dendrites The following code may be helpful.

```
itrig 0 1 dc = 0 pwl (0,0,10us,5n,.2 ms,5n,.21 ms,0)
itrig1 0 21 dc = 0 pwl (0,0,10us,5n,.2 ms,5n,.21 ms,0)
x1 1 2 seg1
x2 2 3 seg1
x3 3 4 seg1
x4 4 5 seg1
*
x21 21 22 seg1
x22 22 23 seg1
x23 23 24 seg1
x24 24 5 seg1
*
x5 5 6 segAND
x6 6 7 seg1
x7 7 8 seg1
x8 8 9 seg1
x9 9 10 seg1
x10 10 11 seg1
Subcircuit segment:
.subckt segAND 2 3
rNernst 30 2 212MEG
vNernst 30 0 -70 mV
cseg 2 0 14p ic = -70 mV
rseg 2 3 100MEG
.ends
```

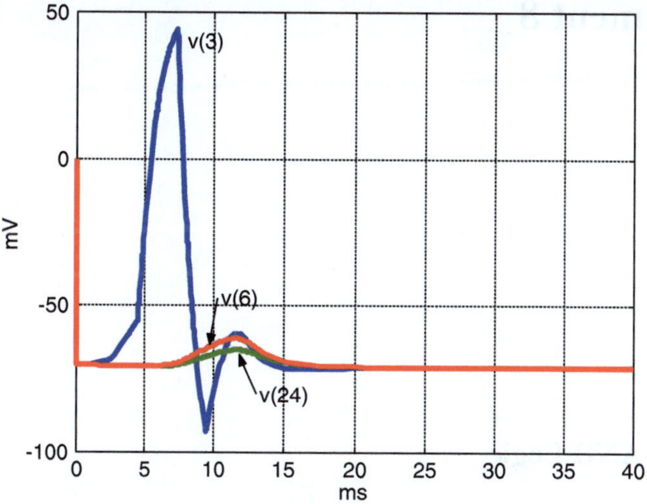

Fig. A.20 Dendritic AND1 using only one input; V(6) shows no output

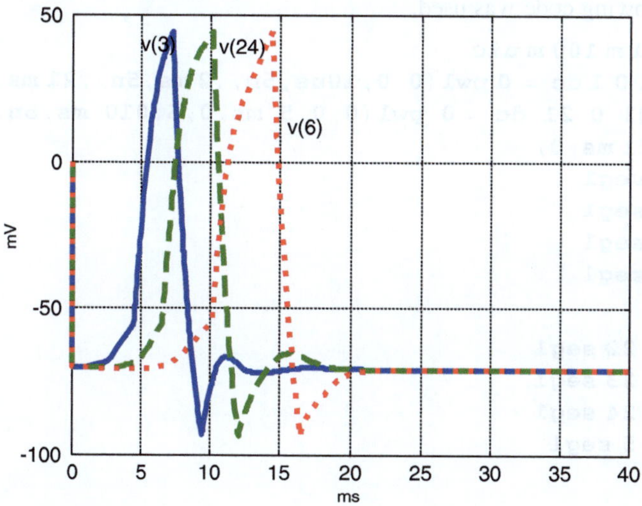

Fig. A.21 Dendritic AND2 using two inputs; V(6) is the output

Fig. A.20 shows that one input does not pass through; Fig. A.21 shows that two inputs will provide output.

- Determine the capacitive parameters on the segAND segment beyond which the AND function does not occur. Explain why not.

Experiment 8

Enabled AND Logic

Purpose: The purpose is to demonstrate AND logic in which charge is stored in capacitance and accumulates to trigger the next segments. One input fails; two inputs even at differing times will trigger an output.

The following code was used.

```
.tran 1 m 100 m uic
itrig1 0 1 dc = 0 pwl (0,0,10us,5n,.2 ms,5n,.21 ms,0)
itrig21 0 21 dc = 0 pwl (0,0,5 ms,0,5.010 ms,5n,5.2 ms,
5n,5.21 ms,0)
x1 1 2 seg1
x2 2 3 seg1
x3 3 4 seg1
x4 4 5 seg1
*
x21 21 22 seg1
x22 22 23 seg1
x23 23 24 seg1
x24 24 5 seg1
*
c1 5 0 100pf ic = -70mv
x5 5 6 seg1
x6 6 7 seg1
x7 7 8 seg1
x8 8 9 seg1
x9 9 10 seg1
x10 10 11 seg1
```

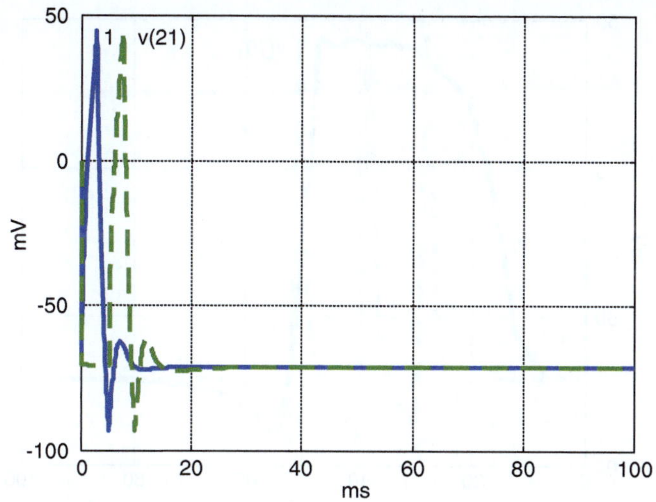

Fig. A.22 Enabled Logic Inputs

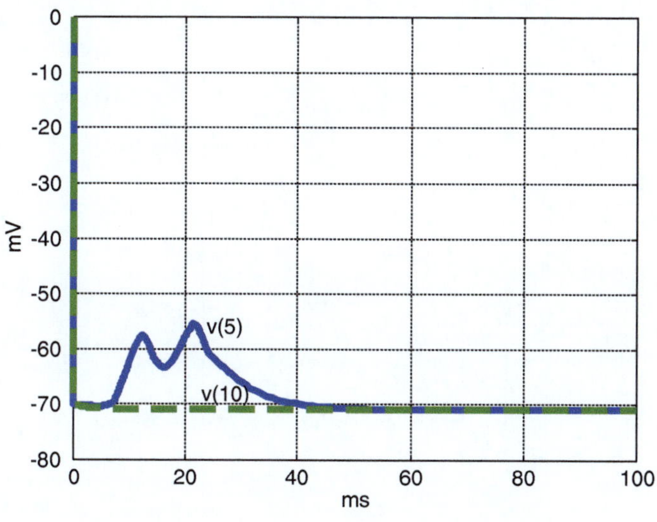

Fig. A.23 One input applied at 10 us, 100 pF at v(5), no output at v(10)

Fig. A.22 illustrates the pulses to be used as inputs. Fig. A.23 shows that one input produces no output (V10). Fig. A.24 shows that two inputs produce an output (V10).

- What is the permitted time spread, approximately, between inputs?
- How much can capacitance C_1 vary and still produce the AND effect?

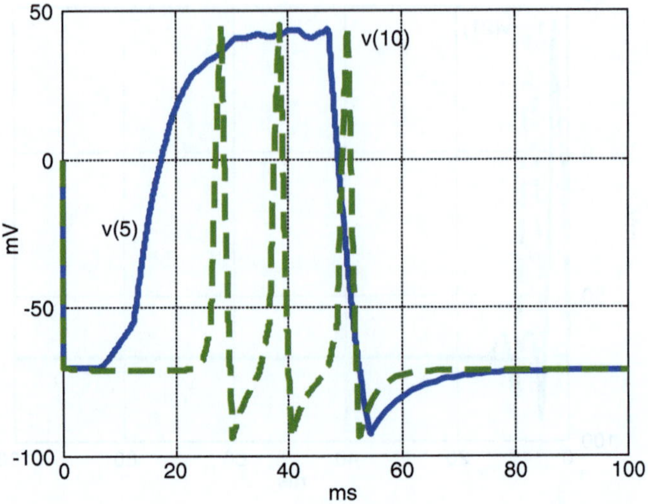

Fig. A.24 Both inputs applied, one at 10 us, one at 5 ms. Output consists of 3 pulses in a burst

Experiment 9

Dendritic XOR Logic

Purpose: The purpose is to demonstrate XOR logic in dendrites

The following code may be helpful.

```
itrig 0 1 dc = 0 pwl(0,0,10us,5n,.2 ms,5n,.21 ms,0)
itrig1 0 21 dc = 0 pwl(0,0,10us,5n,.2 ms,5n,.21 ms,0)
x1 1 2 seg1
x2 2 3 seg1
x3 3 4 seg1
x4 4 5 seg1
*
x21 21 22 seg1
x22 22 23 seg1
x23 23 24 seg1
x24 24 5 seg1
*
x5 5 6 segXOR
x6 6 7 seg1
x7 7 8 seg1
x8 8 9 seg1
x9 9 10 seg1
x10 10 11 seg1
The following segment was used.
```

```
.subckt segXOR 2 3
rNernst 30 2 106MEG
vNernst 30 0 -70 mV
cseg 2 0 20p ic = -70 mV
rseg 2 3 200MEG
.ends
```

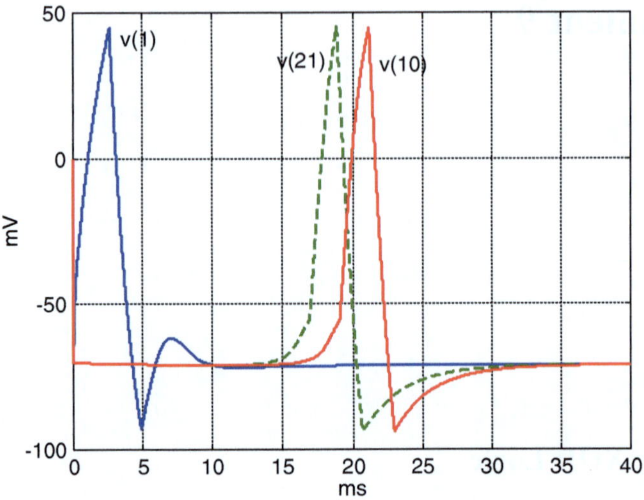

Fig. A.25 Dendritic XOR1 V(1) is reflected to v(21) and also passed to v(10)

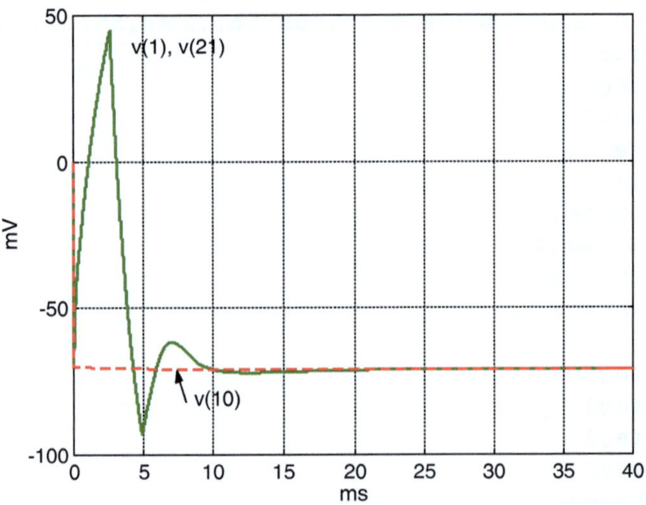

Fig. A.26 Dendritic XOR2 X(1) and x(21) annihilate giving no output at x(10)

Fig. A.25 shows that one input is passed to the output v(10); Fig. A.26 shows that two inputs produce no output at v(10).

- Determine the capacitive parameters on the segXOR segment beyond which the XOR function does not occur. Why do you think it does not occur?

Experiment 10

Multivibration

Purpose: The purpose is to demonstrate multivibration in a neuron

Suggested code:

```
itrig 0 1 dc = 0 pwl (0,0,10us,5n,.2 ms,5n,.21 ms,0)
x1 1 2 seg1
x2 2 3 seg1
x3 3 4 seg1
x4 4 5 seg1
X5 5 6 seg1
x10 6 1 buf
```

Buffer-synapse model:

```
.subckt buf 10 13
vref 1 0 -70 mV
G3 1 13 10 1 5 nA
rload 13 1 400MEG
cload 13 1 2pf ic = −70 m
.ends
```

Fig. A.27 simulates multivibration.

- How can frequency be cut in half?

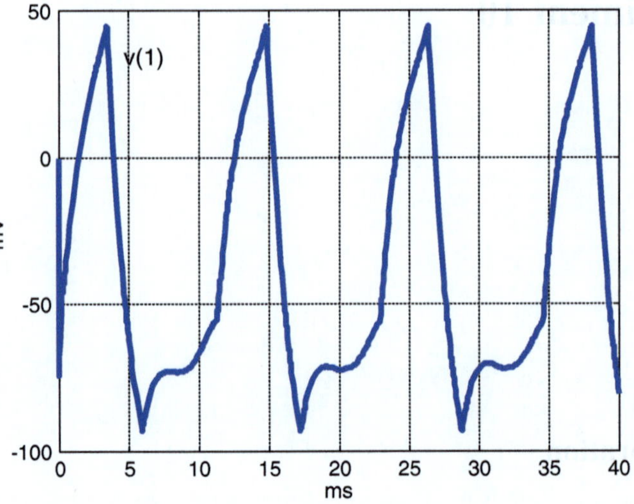

Fig. A.27 Multivibrator MV, 5 segments in a loop closed with a synapse model

Experiment 11

Toggle Using Inhibitory Model

Purpose: The purpose is to demonstrate toggling using negative charge to simulate inhibitory neurotransmitters?

The following code may be helpful.

```
vtrig 22 0 dc = 0 pwl (0,0,10us,1 V,.2 ms,1 V,.21 ms,0)
x21 22 2 bufv
x22 22 5 bufvN
vtrig1 23 0 dc = 0 pwl
(0,0,50 ms,0,50.01 ms,1 V,50.2 ms,1 V,50.21 ms,0)
x23 23 5 bufvN
x24 23 2 bufv
*starts if both x21, x22 applied at t = 0
*stops if both x21, x22 applied at t = 50 ms
x1 2 3 seg1
x2 3 4 seg1
x3 4 5 seg1
*
x4 5 6 seg1
X5 6 7 seg1
*
x10 7 2 buf
```

(See Figs A28–29)

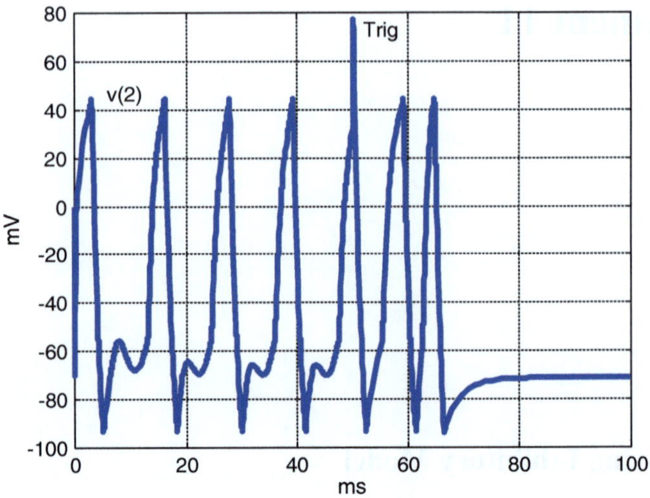

Fig. A.28 Tog Using simultaneous positive & negative charge injection at 0 ms to start, nodes 2 & 5; negative charge injection at 50 ms, nodes 2 & 5 to stop

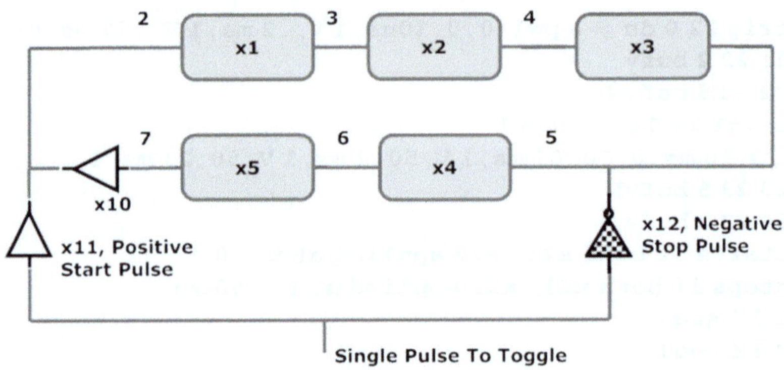

Fig. A.29 Toggle circuit using positive and negative charges

Suggested models for injecting positive or negative charge are as follows.

```
.subckt bufv 10 13
G3 0 13 10 0 10 nA
.ends
*
.subckt bufvN 10 13
G3 13 0 10 0 40 nA
.ends
```

Fig. A.28 simulates the starting of multivibration at 0 ms and the stopping of multivibration at about 60 ms.

- At what current value is the injected negative charge insufficient to stop multivibration?
- Why are there two charge injection points?

Experiment 12

Toggle Using AND/XOR Logic

Purpose: The purpose is to demonstrate neural toggling using AND/XOR Logic
The code used:

```
*itrig 0 1 dc = 0 pwl(0,0,10us,5n,.2 ms,5n,.21 ms,0)
vtrig1 22 0 dc = 0 pwl(0,0,50 ms,0,50.01 ms,1 V,50.2 ms,
1 V,50.21 ms,0)
x1 1 2 seg1
x2 2 3 segXOR
x3 3 4 seg1
x4 4 5 seg1
x5 5 6 seg1
X6 6 7 seg1
*
x10 7 1 buf
x11 22 1 bufv
x12 22 24 bufv
x13 7 24 buf1
x14 24 25 segAND
x15 25 26 buf
x16 26 2 seg1
```

The buffer-synapses used are given below. Buf1 is a simulation artifice that serves to compensate the overly strong x10.

```
.subckt buf 10 13
vref 1 0 -70 mV
G3 0 13 10 1 5 nA
rload 13 1 400MEG
cload 13 1 2pf ic = -70 m
.ends
*
```

(See Figs A30–32)

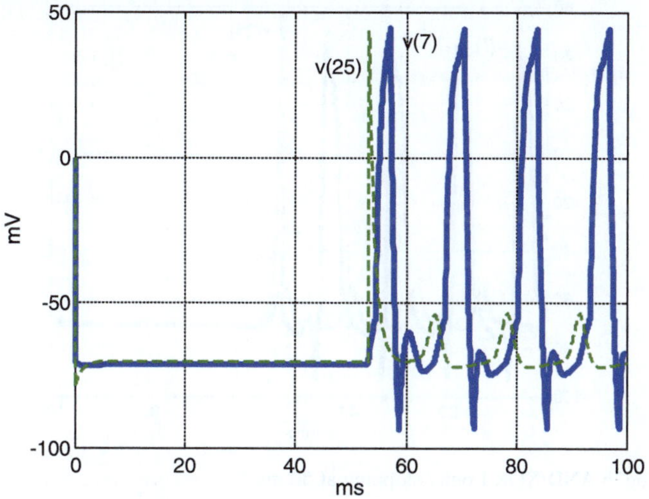

Fig. A.30 Tog2a Using AND/XOR logic/ Starting at 50 ms

```
.subckt bufv 10 13
G3 0 13 10 0 10 nA
.ends
*
.subckt buf1 10 13
vref 1 0 -70 mV
G3 0 13 10 1 3 nA
rload 13 1 100MEG
cload 13 1 2pf ic = −70 m
.ends
```

Fig. A.30 simulates the starting of multivibration; Fig. A.31 simulates the stopping of multivibration.

- Make the "stop" trigger come a few milliseconds sooner, say 3 ms sooner. What is the effect on toggling from true (multivibration) to false (rest)? Explain!

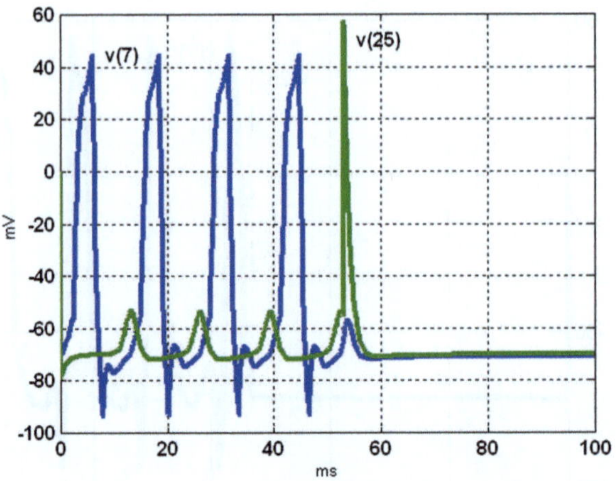

Fig. A.31 Tog2b AND/XOR Logic/ stopping at 50 ms

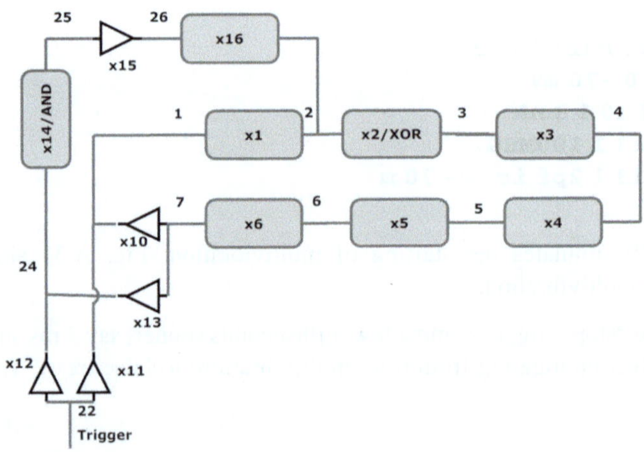

Fig. A.32 AND/XOR diagram of a toggle circuit

Appendix B

Listing of Sample WinSpice Code

*****Single Pulse WinSpice Code Available at: http://winspice.com**

```
.control
destroy all
run
plot v(1)
write SingPul.csv v(1)
.endc
*
.tran 1 m 40 m uic
itrig 0 1 dc = 0 pwl(0,0,10us,5n,.2 ms,5n,.21 ms,0)
x1 1 2 seg1
******************
.subckt seg1 2 3
rNernst 30 2 106MEG
vNernst 30 0 -70 mV
cseg 2 0 31.4p ic = −70 mV
rseg 2 3 200MEG
xcur 2 0 cur1
.ends
******************
.subckt segOR 2 3
rNernst 30 2 106MEG
vNernst 30 0 -70 mV
cseg 2 0 31.4p ic = −70 mV
rseg 2 3 200MEG
*rload 3 0 1000MEG
xcur 2 0 cur1
.ends
```

J.R. Burger, *Brain Theory From A Circuits And Systems Perspective*, Springer Series in Cognitive and Neural Systems 6, DOI 10.1007/978-1-4614-6412-9,
© Springer Science+Business Media New York 2013

```
******************
.subckt segAND 2 3
rNernst 30 2 212MEG
vNernst 30 0 -70 mV
cseg 2 0 14p ic = -70 mV
rseg 2 3 100MEG
.ends
******************
.subckt segXOR 2 3
rNernst 30 2 106MEG
vNernst 30 0 -70 mV
cseg 2 0 20p ic = -70 mV
rseg 2 3 200MEG
xcur 2 0 cur1
.ends
*******************
.subckt segGen2 2 3
vcontrol 30 0 pwl (0 -70mv 60 ms -70 mV 61 ms -45 mV 80 ms -45 mV
81 ms -70 mV)
rNernst 30 2 106MEG
*vNernst 30 0 -45 mV
cseg 2 0 31.4p ic = -70 mV
rseg 2 3 200MEG
xcur 2 0 cur1
.ends
******************
.subckt segGen1 2 3
vcontrol 30 0 pwl (0 -70mv 1 ms -45 mV 40 ms -45 mV 41 ms -70 mV)
rNernst 30 2 106MEG
*vNernst 30 0 -45 mV
cseg 2 0 31.4p ic = -70 mV
rseg 2 3 200MEG
xcur 2 0 cur1
.ends
******************
*Curents/Na loop
.subckt cur1 2 0
vdd 6 0 +300 mV
vss 7 0 -300 mV
VR 44 0 -55 mV
rref 44 4 1MEG
cref 4 0 100p ic = 100 mV
Ri 2 8 1 G
*Rf 8 1 100 G
EXXX 5 0 8 4 100
```

```
cpad 8 0 1f ic = -70 m
rout 5 1 1 k
d1 1 6 diode1
d2 7 1 diode1
.model diode1 D(is = 1.0e-14,cjo = 2p)
vfloat 9 1 1000 mV
rbuf 9 99 1MEG
rL 99 0 100MEG
G1 0 2 99 0 2.11 nA
G2 2 0 199 0 0.955 nA
sNa 99 0 0 1 comp
sNaOFF 99 0 21 0 switch2
sK 199 0 0 11 comp
.model comp sw(ron = 1 roff = 100MEG)
.model switch2 sw(ron = 1 roff = 100MEG)
*****************
* K loop
VR1 43 0 -70 mV
rref1 43 14 1MEG
cref1 14 0 100p ic = 100 mV
Ri1 2 18 10 G
Rf1 18 11 300 G
EXXX1 15 0 18 14 100
cpad1 18 0 1f ic = -70 m
rout1 15 11 1 k
d31 11 6 diode1
d41 7 11 diode1
vfloat1 19 11 1000 mV
rbuf1 19 199 1MEG
rLL1 199 0 100MEG
*****************
*Na OFF loop
VR2 42 0 -5 mV
rref2 42 24 1MEG
cref2 24 0 100p ic = 100 mV
Ri2 2 28 10 G
Rf2 28 21 150 G
EXXX2 25 0 28 24 100
cpad2 28 0 1f ic = -70 m
rout2 25 21 1 k
d51 21 6 diode1
d61 7 21 diode1
vfloat2 29 21 1000 mV
rLL2 29 0 1MEG
.ends
*
```

```
.subckt buf 10 13
vref 1 0 -70 mV
G3 0 13 10 1 5 nA
rload 13 1 400MEG
cload 13 1 2pf ic = −70 m
.ends
*
.subckt bufv 10 13
G3 0 13 10 0 10 nA
.ends
*
.subckt buf1 10 13
vref 1 0 -70 mV
G3 0 13 10 1 3 nA
rload 13 1 100MEG
cload 13 1 2pf ic = −70 m
.ends
*
.end
***☺
```

Index

J.R. Burger, *Brain Theory From A Circuits And Systems Perspective*, Springer Series in Cognitive and Neural Systems 6, DOI 10.1007/978-1-4614-6412-9,
© Springer Science+Business Media New York 2013

MIX
Papier aus verantwortungsvollen Quellen
Paper from responsible sources
FSC® C105338

If you have any concerns about our products,
you can contact us on
ProductSafety@springernature.com

In case Publisher is established outside the EU,
the EU authorized representative is:
Springer Nature Customer Service Center GmbH
Europaplatz 3, 69115 Heidelberg, Germany

Printed by Libri Plureos GmbH
in Hamburg, Germany